ALSO BY SIMON LEVAY

Here Be Dragons:
The Scientific Quest for Extraterrestrial Life
(with David Koerner)

The Earth in Turmoil: Earthquakes and
Volcanoes and Their Impact on Humankind
(with Kerry Sieh)

Albrick's Gold

Queer Science: The Use and Abuse of Research
into Homosexuality

City of Friends: A Portrait of the Gay and Lesbian
Community in America
(with Elisabeth Nonas)

The Sexual Brain

HEALING

THE

BRAIN

HEALING
THE
BRAIN

A Doctor's Controversial Quest for a
Cell Therapy to Cure Parkinson's Disease

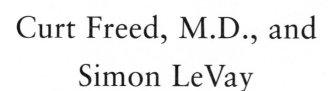

Curt Freed, M.D., and
Simon LeVay

TIMES BOOKS

Henry Holt and Company   New York

Times Books
Henry Holt and Company, LLC
Publishers since 1866
115 West 18th Street
New York, New York 10011

Library of Congress Cataloging-in-Publication Data

Freed, Curt.
  Healing the brain / Curt Freed and Simon LeVay.
    p. cm.
  Includes index.
  ISBN 0-8050-7091-5 (hb)
    1. Parkinson's disease—Treatment—Popular works. 2. Stem
cells. 3. Cellular therapy. I. LeVay, Simon. II. Title.

RC382 .F744 2002
616.8'3306—dc21   2002019671

Henry Holt Books are available for special
promotions and premiums. For details contact:
Director, Special Markets.

First Edition 2002

Designed by Paula Russell Szafranski

Printed in the United States of America

1   3   5   7   9   10   8   6   4   2

HEALING
THE
BRAIN

A Doctor's Controversial Quest for a
Cell Therapy to Cure Parkinson's Disease

~

Curt Freed, M.D., and
Simon LeVay

TIMES BOOKS

Henry Holt and Company   New York

Times Books
Henry Holt and Company, LLC
Publishers since 1866
115 West 18th Street
New York, New York 10011

Library of Congress Cataloging-in-Publication Data

Freed, Curt.
  Healing the brain / Curt Freed and Simon LeVay.
    p. cm.
  Includes index.
  ISBN 0-8050-7091-5 (hb)
    1. Parkinson's disease—Treatment—Popular works. 2. Stem
cells. 3. Cellular therapy. I. LeVay, Simon. II. Title.

RC382 .F744 2002
616.8'3306—dc21   2002019671

Henry Holt Books are available for special
promotions and premiums. For details contact:
Director, Special Markets.

First Edition 2002

Designed by Paula Russell Szafranski

Printed in the United States of America

1 3 5 7 9 10 8 6 4 2

Contents

Contents

Foreword

It may be no coincidence that Dr. Curt Freed's work is performed within sight of the Front Range of the Rockies. In my ten years of working as a patient advocate seeking better funding for research on Parkinson's, I have observed similar traits between the brave souls who first successfully crossed those mountains to reach the Pacific Ocean and the central figures of this book.

This book tells the story of the developing science of brain repair. My introduction to this world began when Dr. Freed and I teamed up in 1991 to educate Congress about Parkinson's in the effort to lift the ban on fetal-cell transplantation research. I was four years into life with a Parkinson's diagnosis and the labels that describe it: "degenerative," "progressive," and worst of all, "incurable." They presented a bleak future if biomedical research did not make a big leap soon.

While I greatly appreciate Dr. Freed's continuing willingness to help educate national leaders about the promise and needs of this research, my deepest gratitude is reserved for his dedication to this infant science. As this book describes, he and his elite fraternity of

international colleagues are participating in a revolutionary transformation in the treatment of brain disorders and doing so against
obstacles comparable to the huge mountains in his backyard.

The first quality this group possesses is vision, to make the big leap
required in the field of brain research. Many brain disorders don't
even have temporarily effective symptomatic relief—Parkinson's
L-dopa is one of the few. Restorative or curative brain treatments,
of the sort comparable to those developed for heart conditions, have
until recently been in the realm of science fiction. It has been easy
to believe the skeptics who discount these explorations as fanciful
because the brain is inherently "different."

But these scientists have persevered, recognizing that there were
naysayers in the early days of heart transplantation, and that whatever the field—from art to the Internet—new approaches that now
seem inevitable and simple were regarded at the beginning as foolish
and impossible. One of Dr. Freed's colleagues responds to the skeptics with a slide of a helicopter: "Who would have thought this
machine could fly? But eventually, it did," he says.

Beyond having a vision, this work requires painstaking effort.
Another colleague of Dr. Freed's describes their work as, predominantly, the process of "going down alleys to establish that they are
blind." It takes a special talent to maintain and reshape a vision
while hitting blind alley after blind alley, inspecting what is learned
at each step for vital clues that will, eventually, shape an effective
therapy.

There are more hurdles. Money for Parkinson's research is
scarce, and so the researchers must also be constant grant writers
and fund-raisers. The financial obstacles have seemed at times more
daunting than the science itself: despite the huge scientific promise,
the field is only now beginning to develop a fraction of the public
and governmental support enjoyed by higher profile diseases like
AIDS and breast cancer.

Because Parkinson's has been the model disease for new "high-
tech" therapies—testing applications using fetal-cell transplants and
stem cells, and evaluating so-called cloned cells—these researchers
have endured presidential and congressional bans and inquisitions,

media scrutiny, picketing on occasion—and uncertainty at every turn.

At times I've been amazed that the field could progress at all under these conditions. But the scientists who have chosen the field have brought all the talents required. Raw intellect, which is present in vast quantities in this club, has been joined by the pioneering spirit that is so evident in Dr. Freed and Simon LeVay's narrative. Each challenge has been assaulted with a combination of vision, scientific expertise, and a refusal to be discouraged.

This book describes the revolutionary changes that have resulted: the brain's plasticity and regenerative capacities have been revealed, producing a developing array of promising, high-tech approaches to treatment. With scientists such as Dr. Freed at the helm, they are moving toward solutions for those of us currently diagnosed with Parkinson's. Their findings will pave the way for the application to a host of other brain afflictions—Alzheimer's, Lou Gehrig's, Huntington's, spinal cord injuries, and more.

As a result, the word "incurable" that labeled Parkinson's when I was diagnosed is still technically true, but it sounds increasingly antiquated in this new, hopeful environment. Increasingly, my fear of my future is replaced by feelings of wonder—at reports like those in these pages, describing patients who have had important functions restored. I have an increasing conviction that this "helicopter" will "fly," and that it will be done in time for many of the millions who are waiting for a rescue.

My ability to be so hopeful is the consequence of the unrelenting explorations of Dr. Freed and his colleagues. I'm so very grateful.

Joan I. Samuelson
President, Parkinson's Action Network

Authors' Note

This book has two authors—Curt Freed, whose search for a cure for Parkinson's disease forms the focus of the book, and Simon LeVay, a science writer with a background in brain research. Because we tell Curt Freed's story, the book is written in the first person singular. In reality, however, *Healing the Brain* is a collaborative effort.

The research described here has engendered a great deal of controversy. The use of tissue derived from aborted human fetuses has raised serious ethical concerns. The "double-blind" neurosurgical trial—in which randomly selected patients were subjected to sham operations—has also been criticized on ethical grounds. The ultimate interpretation of the results is still being energetically debated. We have attempted to present conflicting points of view as fairly as possible, but the main perspective is, of course, that of Curt Freed, the central player in this medical drama.

We owe a debt of gratitude to a number of people who helped us in the preparation of the book. First and foremost, we thank the patients and their spouses who agreed to be interviewed and who gave permission for us to write about them. A few of these individuals asked us not to use their real names: in these cases we have used pseudonyms, and these are indicated by quotation marks where they first appear.

We also thank several colleagues of Curt Freed as well as other professionals who told their sides of the story. These include Robert Breeze, M.D., Stuart Schneck, M.D., Shar Waldrop, Robert Iacono, M.D., Stanley Fahn, M.D., David Eidelberg, M.D., and Vijay Dhawan, Ph.D. Finally, we thank our agent, Julie Castiglia, who played a key role in bringing this project to fruition.

HEALING
THE
BRAIN

CHAPTER 1

Election Day

It was 11 P.M. on Election Day—November 8, 1988—when we wheeled Don Nelson into the operating room. Hardly the perfect time of day to begin a challenging and experimental procedure that would take nine hours to complete—if nothing went wrong. We were all exhausted from a long day of preparations, and we would have liked nothing better than to get a good night's sleep and start the whole thing fresh in the morning. Or to kick back with a few beers and watch the presidential election results come in—an election that I knew would profoundly affect my professional life, for good or bad. But neither of those options was in the cards. It was now or never, and adrenaline would have to carry us through the night.

Don first came to me about a year earlier, but his troubles began way back in 1969, when he was only thirty-two years old. At that time he was the manager of a mobile-home factory, and he lived with his wife, Carolyn, and their two children in Greeley, Colorado. Don has written an account of his illness, so here are some episodes from those early years, in his own words:

In January of 1969 I started to have trouble with certain keys on a manual typewriter I had in my home. The keys used by the little finger of my left hand were not making an imprint. Thinking the typewriter might need adjusting, I took it to a repair shop and was told it was in good condition. I then started to notice a little loss of feeling in my left little finger and also the outside edge of my left hand. I seemed to have trouble picking up a sheet of paper from a flat surface, like the top of my desk.

I was examined by an orthopedic doctor in Greeley who referred me to a hospital in Denver for testing. I was given an electromyogram [electrical recording from muscles] which indicated damage to the ulnar nerve in my left elbow. In June 1969 I had surgery to relocate the nerve. When I regained consciousness my left hand and fingers felt stiff. The surgeon told me that it would be fine once the nerve healed and that it could take six to 12 months. However, it continued to get worse. In about three weeks my hand began to shake. I kept losing strength in my left hand and eventually in my left arm. This arm was not responding to my commands—I had to move it with my right arm. If I walked through a door, quite often I'd get caught because my left hand would not release the doorknob. I also had trouble lifting my hand and turning on the turn signals in my car.

In September 1970 I was given an electroencephalogram [recording of brain activity via electrodes on the scalp], but it did not indicate anything new. Early the following year I had gotten so weak that I could not lift up my left arm, it just felt heavy. And my left leg was getting weak.

In the summer of 1971 I saw an orthopedic surgeon in Michigan who had operated on my knee years before. He said the problem was in my spinal cord in my neck area. He scheduled me for a myelogram [an injection of radio-opaque dye into the space surrounding the spinal cord, followed by an X ray]. The only thing that resulted was a terrific headache that lasted almost four weeks. I could not sit up for more than an hour.

We had moved to Florida in connection with my work, and I went to a neurology clinic in Tallahassee. There I had more electroencephalograms, a carotid angiogram [in which X rays are taken of the head during injection of radio-opaque dye into the carotid artery], and a pneumoencephalogram [in which X rays are taken while the fluid in the cerebral ventricles has been replaced with air]. The tests were all negative, but two of the doctors there felt quite sure that I had a small brain tumor, too small to be seen on X-rays, and most likely inoperable. The third doctor thought I was just acting.

From this point on my mind was out of control. Almost every night my mind was thinking about a doctor with a drill and chisel trying to get into my head to remove a tumor. These nightmares went on for several months. . . .

In the summer of 1972 we returned to Denver. I made an appointment with a neurosurgeon. He reviewed the X-rays from Florida and he thought he could see a very minute spot in my brain. He did another angiogram and pneumoencephalogram, and more electroencephalograms, yet with all these tests he could not locate a tumor. He said, give it time and it will grow and be found. I made up my mind I would not agree to another pneumoencephalogram, as the pain and discomfort were more than I could take.

I told the doctor that something else might be the problem and if it wasn't found and attended to, it might be too late and beyond help. I must have gotten to be a pest, because in May of 1973 he arranged an examination with a neurologist on the staff of the University of Colorado Medical Center, Dr. Stuart Schneck. Dr. Schneck examined me for about an hour: then he told me he was quite sure I had Parkinson's disease.

Stuart Schneck, a good friend and colleague of mine, was right. And that correct diagnosis allowed Don to start on drug treatment that greatly alleviated his symptoms—for a while. Eventually he went downhill again, reaching a much more profound depth of disability, as I'll recount later. That's what brought him to me. That's why he was in the operating room, and why my neurosurgeon colleague Bob Breeze was attaching to Don's shaven head a steel contraption that looked like a medieval instrument of torture. And why I was busying myself with a plastic vial that contained a barely discernible collection of cells—cells that, earlier that day, had been part of the brain of a tiny human fetus.

Meanwhile, in a nearby room, an anxious group of people whose lives had been radically affected by Don's illness was gathered. Foremost among them was Don's wife, Carolyn. With her were their son, Kurt, and daughter, Lori, their daughter-in-law, Connie, and son-in-law, Dan, their granddaughters Melissa, Courtney, and Amanda, and Carolyn's sister Elaine with her husband, Bill. Talking with them earlier that evening, I was made all too aware of how much depended on the success of what we were about to attempt, and how great would be the cost of failure.

I thought I was keeping a calm exterior as our preparations went forward and as Bob Breeze and I reviewed every detail of our plans, but evidently I didn't succeed. "The last few hours before the operation, you could see Curt getting progressively more and more agitated," Bob recollected recently. "This experiment that he had done over and over in rats and in monkeys—it was just sinking in that we were going to do it on a human being. We went through a checklist of things that had to be done, but I could see that the real question in Curt's mind was 'Should we be doing this?' But I came from a totally different perspective. I do things to people's brains all the time. So I did my best to reassure him."

What *I* remember distinctly was Bob's final comment, which was "Shall we go ahead?" I replied: "Let's do it."

CHAPTER 2

The Shaking Palsy

Hoxton Square, in the London parish of Shoreditch, has seen better days. The elegant Georgian terraces that once framed the square have long since fallen victim to decay, to the Blitz, and to commercial development. The present buildings represent an uncomfortable mixture of styles and functions. Number 1, at the square's southwest corner, is a modern brick structure whose ground floor serves as a tapas bar. Only a round blue plaque, set high in the wall, reminds the passerby that this site was once the home of a distinguished personage: James Parkinson.

With an effort of the imagination, we can wish away the parked cars and the electric streetlights—and the gaslights that preceded them—and rebuild the eighteenth-century terraces, the cobbled square, and the high wall that once ringed the central garden. We may even spot the sixty-two-year-old surgeon-apothecary himself, as he returns home from a series of house calls on a chilly damp evening in the winter of 1817. Below average height, and wrapped in a heavy black topcoat, he pauses at the top of the steps to scrape the street ordure from his boots and to regain his breath in the smoky air.

In his earlier years, inspired by the ideals of the French Revolution, James Parkinson was a political radical. He was a member of the London Corresponding Society, a group that agitated for electoral reform. He was even a peripheral character in the "Pop-Gun Plot" of 1794—an alleged conspiracy to assassinate King George III during a theatrical performance.

But now, with increasing age and the changing political climate, Mr. Parkinson has evolved into a pillar of the medical establishment. He is no physician, to be sure—he has no university or medical degree. His skill in the healing arts comes from a long apprenticeship to his father, just as he in turn has apprenticed his own son. But his medical expertise, his long devotion to his practice, and a series of monographs and pamphlets he has written on a variety of topics have together cemented his reputation, and he has just been elected president of the Worshipful Society of Apothecaries.

Mr. Parkinson's house calls have caused dinner to be delayed, we may imagine, so he quickly sits down with his wife, Mary, and their unmarried daughter to eat. Afterward, as his family prepares to retire, Parkinson crosses the yard to a smaller building fronting on a side alley. This is his office, though one might mistake it for a geologist's laboratory: the glass jars of the apothecary's trade are outnumbered by rows of cabinets crammed full with fossils. Parkinson's consuming hobby is "oryctology," or what we now call paleontology. He is the author not only of medical works, but also of a three-volume treatise entitled *Organic Remains of a Former World*.

Parkinson dresses the lamp, fills his inkwell, and prepares to launch into yet another monograph, one that he has long intended to write. For years, he took care of a patient, a gardener by trade, who suffered from an unusual and progressive disorder of movement. Eventually that patient succumbed to his disease. But over the ensuing years Parkinson has come across others with similar symptoms. Most of them Parkinson has encountered in the street;

their labored style of movement, so similar to that of his original patient, drew them to his attention, and he would stop them and ask about their malady. Thus he has come to realize that his patient's history was typical of a hitherto undescribed disorder—the "shaking palsy," as he terms it. It is not a particularly common condition, certainly, but neither is it excessively rare. Parkinson has decided to document it for medical science:

> So slight and nearly imperceptible are the first inroads of this malady, and so extremely slow is its progress, that it rarely happens that the patient can form any recollection of the precise period of its commencement. The first symptoms perceived are a slight sense of weakness, with a proneness to trembling in some particular part; sometimes in the head, but most commonly in one of the hands and arms. These symptoms gradually increase in the part first affected; and at an uncertain period, but seldom in less than twelve months or more, the morbid influence is felt in some other part. Thus assuming one of the hands and arms to be first attacked, the other at this period becomes similarly affected. After a few more months the patient is found to be less strict than usual in preserving an upright posture: this being most observable whilst walking, but sometimes whilst sitting or standing. Sometime after the appearance of this symptom, and during its slow increase, one of the legs is discovered slightly to tremble, and is also found to suffer fatigue sooner than the leg of the other side: and in a few months this limb becomes agitated by similar tremblings, and suffers a similar loss of power.
>
> Hitherto the patient will have experienced but little inconvenience; and befriended by the strong influence of habitual endurance, would perhaps seldom think of his being the subject of disease, except when reminded of it by the unsteadiness of his hand, whilst writing or employing himself in any nicer kind of manipulation.

But as the disease proceeds, similar employments are accomplished with considerable difficulty, the hand failing to answer with exactness to the dictates of the will. Walking becomes a task which cannot be performed without considerable attention. The legs are not raised to that height, or with that promptitude which the will directs, so that the utmost care is necessary to prevent frequent falls.

At this period the patient experiences much inconvenience, which unhappily is found daily to increase. The submission of the limbs to the directions of the will can hardly ever be obtained in the performance of the most ordinary offices of life. The fingers cannot be disposed of in the proper directions, and applied with certainty to any proposed point. As time and the disease proceed, difficulties increase: writing can now be hardly at all accomplished; and reading, from the tremulous motion, is accomplished with some difficulty. Whilst at meals the fork not being duly directed frequently fails to raise the morsel from the plate: which, when seized, is with much difficulty conveyed to the mouth. At this period the patient seldom experiences a suspension of the agitation of the limbs. Commencing, for instance in one arm, the wearisome agitation is borne until beyond sufferance, when by suddenly changing the posture it is for a time stopped in that limb, to commence, generally, in less than a minute in one of the legs, or in the arm of the other side. Harassed by this tormenting round, the patient has recourse to walking, a mode of exercise to which the sufferers from this malady are in general partial; owing to their attention being thereby somewhat diverted from their unpleasant feelings, by the care and exertion required to ensure its safe performance.

But as the malady proceeds, even this temporary mitigation of suffering from the agitation of the limbs is denied. The propensity to lean forward becomes invincible, and the patient is thereby forced to step on the toes and forepart of the feet, whilst the upper part of the body is thrown so far forward as to render it difficult to avoid falling on the face. In some cases, when this state of the malady is attained, the patient can no longer exercise himself by walking in his usual manner, but is thrown on the toes and forepart of the feet; being, at the same time, irresistibly impelled to take much quicker and shorter steps, and thereby to adopt unwillingly a running pace. In some cases it is found necessary entirely to substitute running for walking; since otherwise the patient, on proceeding only a very few paces, would inevitably fall. . . .

As the disease proceeds towards its last stage, the trunk is almost permanently bowed, the muscular power is more decidedly diminished, and the tremulous agitation becomes violent. The patient walks now with great difficulty, and unable to support himself with his stick, he dares not venture on this exercise, unless assisted by an attendant, who walking backwards before him, prevents his falling forwards, by the pressure of his hands against the forepart of his shoulders. His words are now scarcely intelligible; and he is not only no longer able to feed himself, but when the food is conveyed to the mouth, so much are actions of the muscles of the tongue, pharynx, &c., impeded by impaired action and perpetual agitation, that the food is with difficulty retained in the mouth until masticated; and then as difficultly swallowed. Now also, from the same cause, another very unpleasant circumstance occurs: the saliva fails of being directed to the back part of the fauces, and hence is continually draining from the mouth, mixed with the particles of food, which he is no longer able to clear from the inside of the mouth.

As the debility increases and the influence of the will over the muscles fades away, the tremulous agitation becomes more vehement. It now seldom leaves him for a moment; but even when exhausted nature seizes a small portion of sleep, the motion becomes so violent as not only to shake the bed-hangings, but even the floor and sashes of the room. The chin is now almost immovably bent down upon the sternum. The slops with which he is attempted to be fed, with the saliva, are continually trickling from the mouth. The power of articulation is lost. The urine and faeces are passed involuntarily; and at the last, constant sleepiness, with slight delirium, and other marks of extreme exhaustion, announce the wished-for release.

Even with the benefit of nearly two centuries of medical progress, I am amazed by the accuracy and clarity of his description, based as it was on only one well-studied case and a handful of brief encounters. In the years since, some details have been corrected. He erred in stating that the tremor persists during sleep—it does not. Also, his emphasis on muscular weakness is misplaced. A person with Parkinson's disease has difficulty initiating movements, and the movements that he or she does perform are slow, but that is not because the muscles lack strength. On the contrary, if one attempts to flex a patient's joints, such as the wrist, the muscles resist bending with considerable vigor, giving an impression of rigidity. When the muscles finally yield, they do so in a jerky fashion. This "cogwheel rigidity" is typical of the disease and distinguishes it from disease of true muscular weakness such as muscular dystrophy.

Another striking aspect of the disease that escapes Parkinson's attention is a lack of facial expression. So little do many patients exercise their facial muscles that the skin creases engendered by their use—the smile lines, frown lines, and so on—are all but absent, especially in persons who develop the disease early in life. Thus their faces, though immobile, often have an uncannily youthful appearance.

The untreated disease is potentially fatal, as Parkinson describes it. Yet because the disease predominantly affects the elderly, and because its course is generally protracted over many years or decades, only a minority of those who are diagnosed with the disorder actually die of it. Some other of humanity's numberless afflictions usually ends their lives before this one does. Yet the debility of Parkinson's disease increases the patient's susceptibility to those other afflictions. The tendency to fall, for example, increases the likelihood of hip fractures and head injuries. Difficulty in swallowing can cause food to enter the windpipe, which may cause pneumonia. Thus the overall life expectancy of persons with Parkinson's disease is significantly shortened.

The bells of St. Leonard's parish church—where Parkinson has worshiped for sixty years, and where in a few more he will be buried—strike midnight. But James Parkinson is a night owl, so he sharpens a fresh supply of pens and soldiers on. He describes the individual case histories, and reviews the various symptoms of the disease. He considers the "differential diagnosis"—how this disease is to be distinguished from others. And he speculates at length about the disease's cause. Most likely, he suggests, it results from an inflammation of the spinal cord in the region of the neck. How this inflammation comes about he does not know, and neither do his patients. "Whilst one has attributed this affliction to indulgence of spirituous liquors," he writes, "and another to long lying on the damp ground; the others have been unable to suggest any circumstance whatever, which, in their opinion, could be considered as having given origin, or disposed, to the calamity under which they suffered."

As for the treatment of the disease, Parkinson suggests that ancient medical panacea: phlebotomy. Blood should be drawn from the veins of the neck, he writes, and this should be followed by the application of blistering agents, so as to provoke a purulent discharge. Alternatively, the neck should be incised on both sides, and the incisions kept open with pieces of cork, so as to allow the inflammatory exudates to escape. (With these options for treatment, I have to sympathize with one of Parkinson's patients, who "being fully assured of the incurable nature of his condition, declined making any attempts for relief.")

The hours pass. As Parkinson reaches his concluding paragraphs, we may perhaps hear hoof-steps and the trundling of carts along Old Street, as market-gardeners haul a new day's supply of produce into the city. Perhaps Parkinson will snatch an hour or two of sleep. But before closing his manuscript, he adds a final thought. The real cause of the disease, he writes, will only be discovered by the post-mortem examination of persons who have succumbed to it. And as if to bring that day closer, Parkinson ends with a word of praise for the stouthearted souls who conduct autopsies. "Little is the public aware," he writes, "of the obligations it owes to those who, led by professional ardour, and the dictates of duty, have devoted themselves to these pursuits, under circumstances most unpleasant and forbidding."

CHAPTER 3

The Light That Failed

Autopsies did indeed provide the answer, though it took far longer than Parkinson could have imagined. For several decades after he wrote *The Shaking Palsy*, the monograph was completely ignored. During this period Parkinson was remembered more for his contributions to geology than to medicine. Then, in the 1870s, Jean-Martin Charcot, the great French clinician who is considered the founder of modern neurology, saw more patients with the disorder. He named it la maladie de Parkinson, thus immortalizing his predecessor.

At about the same time, improved methods were developed for preserving, slicing, and staining brain tissue obtained at autopsy. Occasional reports began to appear on the topic of the brains of persons who had died of Parkinson's disease. In 1917 the Austrian neurologist Constantin von Economo reported that the *substantia nigra* (literally the "black substance"—a region of the midbrain containing darkly pigmented neurons) was damaged in some patients who had died of viral encephalitis; these patients had shown Parkinson-like symptoms in the course of their illness. Later, pathologists reported similar findings in patients who had died of the ordinary form of Parkinson's disease. The definitive study was done in 1938 by a German neuroanatomist, Rolf Hassler. Hassler observed that the darkly pigmented nerve cells that give the substantia nigra its characteristic black color were greatly reduced in number in Parkinson's patients, sometimes to such a degree that the substantia nigra could barely be recognized with the naked eye.

Important though Hassler's observation was, it also was neglected for years. This was partly on account of the disruptions caused by the Second World War. And although Hassler continued doing research after the war, he branched off into other topics, including an ethically questionable set of studies in which he destroyed parts of the brains of sex offenders.

It remained for another researcher, Oleh Hornykiewicz of the University of Vienna, to make the breakthrough that revolutionized the treatment of Parkinson's disease. Hornykiewicz is a brilliant and charming man who, now well into his seventies, still runs an active

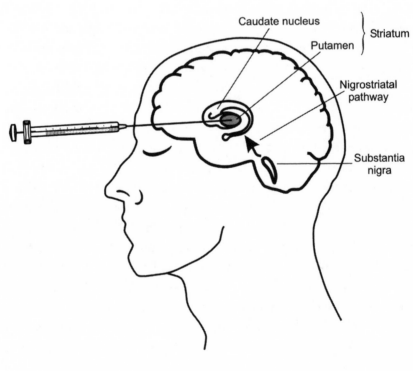

laboratory. He has published memoirs of his life and work that are penned in a delightful *faux*-Proustian style ("To the surface of my consciousness also rise, like snapshots, memories of the hot summers spent in the bountiful countryside . . .").

Oleh's idyllic childhood was interrupted by the outbreak of the Second World War. His family was displaced from their native Galicia (now part of Ukraine) and ended up in Vienna. Too young to participate in a conflict that claimed the lives of several family members, he continued his schooling. After the war ended, he enrolled in medical school. In 1956 he spent a postdoctoral year at the University of Oxford, where his advisor, pharmacologist Hermann Blaschko, set him to studying the properties and function of a newly identified brain chemical, *dopamine*.

Dopamine had been given that name only five years earlier—the word is an easily pronounced selection of letters from the compound's full chemical title: β-3,4-dihydroxyphenylethylamine. It is a member of a family of compounds known as catecholamines, all of which are synthesized from tyrosine, an amino acid available in food. But in the 1950s dopamine was very much the poor relation in the catecholamine family, completely overshadowed by two other compounds, *adrenaline* and *noradrenaline*, which had been discovered in 1910 by Sir Henry Dale. He found these chemicals in high concentrations in the adrenal gland, hence their names. He showed that both adrenaline and noradrenaline could speed up the heart rate and increase blood pressure in animals. Harvard physiologist Walter Cannon showed in the 1930s that the adrenal gland released these compounds in the "fight or flight" response. Noradrenaline, though not familiar to the public, had already been established as a neurotransmitter (a signaling molecule released by nerve cells) in the sympathetic nervous system, which also plays a role in fight-or-flight responses. Dopamine, on the other hand, was dismissed as a mere intermediary—a metabolic step in the synthesis of noradrenaline and adrenaline.

Blaschko suspected that dopamine had some physiological function of its own, however, and Hornykiewicz did indeed demonstrate one such function, albeit a fairly unexciting one: he showed that it helped regulate blood pressure in the guinea pig. Just as he was wrapping up his studies in Oxford, a Swedish group led by Arvid Carlsson reported that dopamine was present in surprisingly high concentrations in the brain itself, both in laboratory animals and in humans. Clearly dopamine had a major role in brain function.

Then, in early 1959, two Swedish workers, Åke Bertler and Evald Rosengren, reported that most of the brain's dopamine was found in a large structure hidden deep beneath the folds of the cerebral cortex, the *striatum*. This name is a shortening of the structure's full title, *corpus striatum* or "striped body"—so-called because bundles of fibers course across it. As with most brain structures, there are two striatums (or striata), one on the left and one on the right side of the brain.

The striatum was known to be involved in the control of movement. Furthermore, Carlsson's studies had shown that a certain drug, reserpine, lowers the levels of dopamine in the brains of experimental animals, leaving them in a somewhat Parkinson-like state of decreased mobility. Thus Bertler and Rosengren suggested that dopamine was necessary for the normal production and control of movements.

The Swedes' paper electrified Hornykiewicz. Could a loss of dopamine in the striatum lie at the root of Parkinson's disease? He raced into action. With his postdoctoral student Herbert Ehringer, Hornykiewicz set out to collect brain tissue from patients who had died of Parkinson's disease, along with tissue from patients who had died of other diseases to serve as controls. They dissected out blocks of tissue from the striatum, ground up the tissue, and carried out a chemical assay for the presence of dopamine.

The Swedish workers had been able to use a new and highly sensitive method for the detection of dopamine, which involved its conversion to a fluorescent compound, followed by the measurement of the fluorescence in a special apparatus. Hornykiewicz and Ehringer lacked this apparatus, however, so they had to use an older and less sensitive technique in which the dopamine was treated with iodine to produce a pink compound, whose concentration was then measured in a "colorimeter." When they took tissue from the striatum of the control brains, the iodine reaction generated a bright pink coloration in the reaction vial that could easily be seen with the naked eye. But when Hornykiewicz and Ehringer carried out the procedure on their first Parkinson's brain, the result could not be more different: the vial remained almost colorless.

It was a thrilling moment for the two men—one of those major discoveries that all scientists hope to make but only a few actually do. Their experiment had not just exposed the key feature of Parkinson's disease, it had also, for virtually the first time, demonstrated the relevance of basic neuroscience to medicine. Indeed, the very notion that diseases had discoverable causes at the molecular level was still in its infancy—only a very few diseases, such as sickle-cell anemia, were understood at that level.

When their excitement subsided, Hornykiewicz and Ehringer realized that they had much work to do. They studied two more Parkinson's brains and then, at the urging of their "Institutsdirektor," studied three more. All six brains showed the same thing: a profound loss of dopamine in the striatum. It was more than a year later, in the fall of 1960, that they sent a written report of their discovery to a scientific journal for publication. It was accepted immediately.

A few weeks later, while Hornykiewicz was correcting the proofs of the manuscript, an even more exciting idea occurred to him. Perhaps, he reasoned, it might be possible to alleviate the symptoms of Parkinson's disease by replacing the missing dopamine. Certainly one could not administer dopamine itself—the molecule was too unstable, it would have major undesirable side effects, and it would not even pass from the blood into the brain. But Hornykiewicz's idea was to administer the chemical precursor of dopamine, a substance called *dihydroxyphenylalanine*, or dopa for short. Dopa is normally synthesized in the body from tyrosine, and dopa in turn is converted into dopamine by the nerve cells that use dopamine as a transmitter. Those nerve cells contain a special enzyme that carries out the conversion; it is named *dopa decarboxylase* to indicate that it clips a "carboxyl" (-COOH) group off the dopa molecule.

Simple as this idea was in principle, there were major practical difficulties. For one thing, dopa comes in two mirror-image forms. Only one of these, the "left-handed" or *L* form of the molecule, could be used by the body to make dopamine. Separating the two forms from each other was no easy task. Hornykiewicz, however, owned a precious two grams of purified L-dopa that had been given to him by the Hoffman-LaRoche pharmaceutical company. He decided to use it all on a do-or-die clinical experiment.

To that end, he hastily contacted an acquaintance of his, Walther Birkmayer, who was in charge of the neurology ward in a home for the aged run by the city of Vienna. A number of victims of Parkinson's disease were housed there. Hornykiewicz explained his theory to Birkmayer and gave him the L-dopa along with detailed instructions as to how to administer it. Birkmayer agreed to test it on some of his patients.

For six months Birkmayer did not test the drug on anyone. Today, such delays—caused by grants officers, human-subject committees, and the like—are the price we accept to ensure the safety of medical research. But Birkmayer was subject to no such oversight. He simply left the drug in his refrigerator and went about his business, leaving Hornykiewicz to fire off increasingly intemperate memos. It turned out that two factors motivated Birkmayer's delay. First, like most people at that time, he thought that dopamine was a physiologically insignificant molecule that was unlikely to have any clinical effects. And second, as he later admitted, he held off on testing the drug to pay Hornykiewicz back for a perceived slight, namely his refusal to participate in an experiment that Birkmayer had dreamed up some years previously.

Finally, in July of 1961, Birkmayer did test the drug on several of his patients with Parkinson's disease. The effect was spectacular. As Birkmayer and Hornykiewicz wrote in the 1961 paper that described their experiments:

> Bedridden patients who were unable to sit up, patients who could not stand up from a sitting position, and patients who, when standing, could not start walking, performed all these activities after L-dopa with ease.

They walked around with normal associated movements and they could even run and jump. The voiceless, aphonic speech, blurred by palilalia [repetition of syllables] and unclear articulation, became forceful and clear again like in a normal person. For short periods of time the patients were able to perform motor activities which could not be prompted to any comparable degree by any other known drug. This dopa-effect reached its peak within 2 to 3 hours and lasted, in diminishing intensity, for 24 hours.

Nowadays we rarely get to witness the power of L-dopa to "resurrect" men and women locked in an almost deathlike state by their disease. That's because most people with Parkinson's are treated with L-dopa from an early stage in their illness; they are not permitted to descend into the locked-in state that was sometimes their fate in earlier times. Yet for the first generation of physicians who used L-dopa, the experience was indelible. Neurologist Oliver Sacks, who tested L-dopa on patients who had a condition resembling Parkinson's disease in New York in 1969, provided a vivid and moving account of the results in his 1973 book, *Awakenings*.

Even with Hornykiewicz and Birkmayer's dramatic report, and a similar one from Andre Barbeau in Canada, L-dopa treatment did not catch on for several years. This was partly because of the difficulty in preparing the drug, and partly the result of a continuing skepticism in the neurological community about dopamine's role in the brain. In spite of the lack of clinical interest, important further discoveries were made about the basic biology of the system. In particular, a link was forged between the two known facts about the Parkinsonian brain: the dopamine deficit in the striatum and the loss of pigmented cells in the substantia nigra.

In 1963, Hornykiewicz showed that dopamine was normally present in the substantia nigra, too, and that dopamine levels were abnormally low in the substantia nigra of Parkinson's patients, just as they were in the striatum. Could there be some connection between the substantia nigra and the striatum? Traditional anatomists, including Hassler, had long denied the existence of a neural pathway linking the two structures. But two Swedish workers, Kjelle Fuxe and Annica Dahlström, applied a recently developed technique for visualizing dopamine in thin slices of brain tissue. This technique involved the conversion of dopamine into a fluorescent compound that could be seen in a fluorescence microscope. They discovered that in the substantia nigra the dopamine was contained within the darkly pigmented cell bodies that gave the substantia nigra its name. In the striatum, on the other hand, the dopamine was located not in neuronal cell bodies but in nerve terminals.

It seemed likely, then, that the cells of the substantia nigra sent fibers (also called *axons*) to the striatum, and that dopamine was present both in the cell bodies and in the axon terminals. The clincher came in 1965, when Louis Poirier and Ted Sourkes of McGill University in Montreal destroyed parts of the substantia nigra in monkeys. The monkeys developed symptoms of Parkinson's disease, and examination of their brains showed that dopamine levels in the striatum had declined drastically. Thus, the existence of nerve fibers running from the substantia nigra to the striatum, and their importance in the symptomatology of the disease, was verified. This *nigrostriatal pathway* has been at the center of all subsequent research into Parkinson's disease.

In 1967, chemist George Cotzias and his colleagues at the Brookhaven National Laboratory in New York demonstrated beyond all doubt that dopa was an effective drug in the treatment of Parkinson's disease. They used the unpurified mixture of D-dopa and L-dopa, but by gradually increasing the dose to levels much higher than those used by Hornykiewicz and Birkmayer, they were able to achieve long-lasting relief of the symptoms of the disease. As Cotzias recognized at the time, giving the D-form of dopa was risky: it was associated with potentially life-threatening anemia.

The publication of Cotzias's paper stimulated the Hoffman-LaRoche company to synthesize large amounts of the purified L-dopa. The company then organized a multicenter trial of the

drug, with extremely good results. One of the doctors who participated in this trial was Stuart Schneck, a Denver-area neurologist
who later became a close collaborator of mine (he has recently
retired from practice). Stuart became familiar with Parkinson's disease as a child because two of his uncles suffered from it. He was
well aware how ineffective the traditional medications had been, so
he jumped at the opportunity to test the new drug. "It was a thrilling experience," he says, "to see people who were wheelchair-
bound get up and walk, to hear people who hadn't been able to
communicate speak intelligibly once more."

Word of the almost magical powers of L-dopa soon entered
popular culture. For one thing, the idea developed that it was a
powerful aphrodisiac—men who took it, once out of their wheel-
chairs, made a beeline for the nearest lady, and vice versa. I'm not
sure that the drug has any effects of this kind, but its reputation
was strong enough to provoke titling a sexually provocative film
L-dopa, shown in theaters in 1971.

One further technical advance cemented L-dopa's central role in
the treatment of Parkinson's disease. As L-dopa is absorbed from
the intestine, enters the bloodstream, and passes through the liver,
much of it is converted to dopamine by the same decarboxylase
enzyme that is present in the brain. Unfortunately, the dopamine
circulating in the bloodstream cannot cross the "blood-brain barrier" into the brain, so it is useless to the Parkinson patient. Thus,
over 90 percent of the L-dopa is wasted. In addition, the circulating dopamine produces troublesome side effects, most especially
chronic nausea. The solution to this problem came with the discovery of another drug, carbidopa. This drug blocks the decarboxylase
enzyme in the liver, but it cannot enter the brain so has no effect
on the decarboxylase enzyme there. When L-dopa and carbidopa
are given in combination, the L-dopa passes through the liver
unchanged, so that much more of it is available to enter the brain
and form dopamine in the striatum. This means that the dose of
L-dopa can be greatly reduced—from about 10 grams per day to
about 1 gram per day—and this in turn means a reduction in the
troublesome side effects of L-dopa treatment, especially nausea. The

L-dopa–carbidopa combination, sold under the trade name of Sinemet, rapidly became the mainstay of treatment for Parkinson's disease. Today, the majority of all people with Parkinson's disease take Sinemet, either alone or in combination with other drugs. In fact, if a patient with symptoms of Parkinson's disease does not respond to L-dopa with an improved ability to move, he or she probably does not have Parkinson's disease.

In spite of its undoubted status as the most important treatment for Parkinson's disease, L-dopa has not put an end to the suffering the disease causes. The drug does nothing to slow the underlying process of Parkinson's disease: it doesn't halt the gradual degeneration and disappearance of the dopamine cells in the substantia nigra and their terminals in the striatum. Yet dopamine terminals must be present for L-dopa to exert its beneficial effect, for it is the terminals that take up the drug, convert it to dopamine, and release the dopamine in a physiologically appropriate fashion.

In some patients, such as those described by Oliver Sacks in *Awakenings*, the drug offers limited benefits even from the start, quite unlike the improvement that Hornykiewicz and Cotzias saw. But Sacks's patients had a special variety of Parkinson's disease: they developed Parkinsonian symptoms as a complication of viral encephalitis (the disease studied by von Economo in Austria in the early twentieth century). While the dopamine cells were indeed damaged in those patients, other areas of their brains were probably abnormal, too. The more common form of the disease, as described by Parkinson himself, is slow in onset and its ultimate cause is still a puzzle. In these "regular" Parkinson's disease patients, L-dopa usually works well for years, but eventually it becomes less and less effective. It's probably the progression of the underlying disease that is responsible for the ultimate "failure" of L-dopa in many patients.

The major ill effect of long-term L-dopa treatment is the appearance of involuntary movements, or *dyskinesias*. These movements are quite different from the involuntary tremor that is so characteristic of the disease. Whereas the Parkinsonian tremor is a steady, oscillating motion of some portion of the body—often a hand—

dyskinesias are irregular writhing or flailing motions that can involve any body part. Sometimes they are accompanied by painful, cramplike contractions of muscle groups, such as the muscles of the neck (*dystonias*).

Dyskinesias and dystonias can become severely disabling: they may directly interfere with the activities of daily living, such as dressing and eating, and they may also aggravate wear-and-tear injuries of joints, ligaments, and muscles. Back pain and disk disease often complicate Parkinson's disease. And though the mind is usually intact, the abnormal movements make patients appear disabled, so that normal social interactions are strained. Jobs are lost prematurely, marriages are broken, and patients may be reluctant to go out in public, sometimes becoming reclusive. The resulting isolation contributes to depression.

Dyskinesias are not a part of the natural history of Parkinson's disease. Rather, they are caused by the very drug that is intended to alleviate the disease's effects. Early in the disease's course, the appearance of dyskinesias is an indication that the patient is being overdosed with L-dopa. A lowering of the dose, or an adjustment of the timing of doses during the day, may alleviate the problem. As the disease progresses, however, and the number of dopamine terminals in the striatum declines, it becomes harder and harder to treat its symptoms effectively without provoking dyskinesias. Eventually, the patient may alternate between two unwelcome states: an underdosed condition, resembling the untreated disease, in which he or she is stiff and immobile, and an overdosed condition characterized by dyskinetic movements. To an onlooker, the underdosed, or "off," condition may seem preferable, but patients themselves often prefer the "on" state, even when it is marred by incessant dyskinesias, because in that condition they at least have some freedom of action.

The reason for these dramatic shifts is still being debated by experts. It's my opinion that the fluctuations are caused by the progression of the disease, not by the drug alone. Because there are so few dopamine nerve terminals remaining in the brain, nerve cells can no longer regulate the amount of dopamine needed for normal movement. When a surge of dopamine comes from a dose of L-dopa, the brain forces the body to move, even with wild swings of the arms, legs, and head.

Since the initial introduction of L-dopa and Sinemet, newer drugs have helped control these problems. A slow-release form of Sinemet can smooth out the "on-off" fluctuations. A class of drugs called *dopamine agonists*, such as pergolide (Permax) and pramipexole (Mirapex), directly mimic dopamine's actions and therefore do not need to be converted into dopamine by nerve terminals in the striatum. These drugs are very helpful for some patients.

Yet a patient who has the disease for more than a decade is likely to become seriously disabled by the disease, even with the best drug treatments available. Perhaps future drug development will improve these prospects. Still, I am convinced that no drug therapy based on dopamine replacement can provide an adequate symptomatic treatment for Parkinson's disease once "on-off" fluctuations have become severe. We already have the best drug—L-dopa. The only better drug would be one that treated the underlying cause of the disease. Several such drugs have been proposed and tested (vitamin E and the monoamine oxidase inhibitor selegilene) but none has proven effective. Since we don't know the cause of the disease in most patients (and the cause may prove to be different in different people), it is hard to search for such a remedy. Instead of looking for new drugs, therefore, my research team has focused on efforts to repair the brain. To explain how I decided to go down this road, I need to recount some of my own scientific career.

In 1961, when Hornykiewicz and Birkmayer first tested L-dopa on patients with Parkinson's disease, I was a freshman at Harvard College. I planned to go to medical school, but I didn't have any clear idea of what branch of medicine I wanted to pursue. Hornykiewicz's work didn't influence me at that time: in fact, I didn't hear of it until years later.

It was in 1965, as I was about to enter Harvard Medical School, that I made up my mind to do brain research. And it wasn't Hornykiewicz's research that motivated me. Instead, it was the work of David Hubel and Torsten Wiesel, scientists at Harvard. Hubel and Wiesel, who later won the Nobel Prize for their work, analyzed the function of the part of the cerebral cortex that deals with vision. I read about their work in a 1965 article they wrote for *Scientific American*. I was struck by the elegance of their research. They showed how vision worked. Their analysis began with understanding that the retina in the eye does not identify objects but instead reports simple patterns of light and dark. Hubel and Wiesel demonstrated how these simple patterns are then assembled and interpreted by carefully organized layers of the visual cortex at the back of the brain. Studying brain function looked like the perfect career challenge—difficult but not impossible. I was certainly not the only young person who was motivated to enter brain science by Hubel and Wiesel's work. My coauthor Simon LeVay, who was then a graduate student in Germany, actually went to spend a year's postdoctoral fellowship in their research group—and ended up staying for thirteen years.

At Harvard Medical School, the basic scientists in the neurobiology department took their teaching seriously. The department's chairman, Steve Kuffler, had assembled a world-class faculty that included Hubel and Wiesel. For every lecture, the entire departmental faculty lined the back row of the lecture hall, supporting and probably critiquing the lecturer for the day. We had some lectures from neurologists; one whom I especially remember was David Poskanzer. He told us that the 1921 influenza epidemic was the cause of nearly every case of Parkinson's disease, so that the disease would disappear when the last patient who lived through that era had died. Unfortunately, he was proved wrong.

The Vietnam War dominated my college and medical school years. Like nearly all of my colleagues, I could not believe that our government could pursue such a pointless and catastrophic conflict. At the same time Lyndon Johnson was revolutionizing American society with civil rights legislation and Medicare, he was waging an expanded war on a third world country with no point whatsoever.

Smart people like Robert McNamara and McGeorge Bundy participated in the insanity, only to later apologize. Worst of all, we who were of an age to participate in the war were being used as a rationale for it. "Support our boys in Vietnam" became the slogan that justified sending and losing more soldiers in that morass. With the election of Richard Nixon and the appointment of Henry Kissinger to the government, lying and brutality intensified. Hundreds of young Americans were being killed every week.

I was so disgusted with the war and our government that I went to England in April 1969 to see if I would prefer living away from the United States. My reason for being in London was to work on a research project to use computers to analyze images of human chromosomes. I had been working with computers since 1961, and on image analysis since 1965. My weeks in London helped in the research, but my experience there also made me realize that I was absolutely an American who needed to fight the Nixon administration and combat the damage it was doing to our society. I vowed to do everything possible to change the disastrous American policy.

The experiences of the 1960s were a major influence on my political and scientific decisions thereafter. With Vietnam, it had become clear we could not assume the government was either right or reasonable. We as citizens had to speak up to change policies. What our country offered was the chance to argue for those changes and to work to replace politicians and to replace bad policies.

After medical school I headed to Los Angeles to begin a residency in internal medicine in the Harbor General Hospital/UCLA program. In those two years, we all changed from being student observers to action-oriented physicians. The key to that transformation was learning to make decisions with almost any information base. Because many of the county hospital patients arrived desperately ill, we had a term called "pre-arrest resuscitation." Recognizing that something was catastrophically wrong and needed to be fixed *before* the patient suffered a cardiac arrest was key to saving lives. Recognizing what was treatable and was not treatable was the other lesson.

I continued to develop a perspective on brain-body interaction with a year's residency in psychiatry at Massachusetts General Hospital in Boston. As with internal medicine, psychiatry proved to be a revelation in terms of understanding brain function and seeing how effective drugs could be at controlling signs of mental illness. Schizophrenia and manic depressive disease are distinct illnesses, and patients with these disorders fit into categories just like patients with heart failure or asthma. The "medical model" and drug treatment apply equally well to physical and psychiatric illness.

To pursue the key role of drugs in managing most diseases, in 1972 I headed to the University of California, San Francisco, for a three-year postdoctoral fellowship in clinical pharmacology with Ken Melmon and Neal Castagnoli. My goal was to study the differences between two closely related drugs that worked on the brain. The first was L-dopa, which is used almost exclusively to treat patients with Parkinson's disease. The second was α-methyldopa, a drug used to treat high blood pressure. In 1975, after completing this project, I moved to the University of Colorado as assistant professor of medicine and pharmacology.

The details of my early research would doubtless strike most readers as arcane and tedious: developing high-sensitivity assays for dopamine and its breakdown products, for example, and tracing the metabolic pathways that the body used to synthesize and destroy dopamine and other catecholamines. But I should describe two kinds of experiments that really influenced my thinking about the treatment of Parkinson's disease.

In one set of experiments, which I did in collaboration with Bob Murphy, Peg Hoehn, and Tom Crowley here at the University of Colorado, we made a close study of the effects of L-dopa on patients with Parkinson's disease. We gave the volunteers their L-dopa dose every two and a half hours, but the dose could be either the real drug or a placebo (an inert pill resembling the L-dopa pill). The idea here was to control for effects that might be due solely to the patient's expectation that he or she would improve after taking the pill. Then, every thirty minutes, we took blood samples. Using the sensitive assay procedures that we had developed, we measured the levels of L-dopa in the blood samples. In addition, every sixty minutes we performed a neurological examination to see exactly how impaired the patient's movements were. When we analyzed our data we found that the patients' clinical state was tied very closely to the levels of L-dopa in the blood at that time: as L-dopa levels rose after taking the drug, the patients' movements improved, and as the L-dopa levels declined (when the patient was getting the placebo, for example) his movements worsened. A patient with Parkinson's disease, in other words, is a slave to his drug levels on an almost minute-by-minute basis.

Why was timing so important? To study this question, we did a set of experiments in rats. My postdoc, Bryon Yamamoto, put electrodes into the left and right striatum of the rats. These electrodes had a special surface at their tips that was sensitive to dopamine: in the presence of dopamine, a small current would flow, which we could measure. We trained the rats to make a continuous turning motion, as if chasing their own tails: cued by us, the rat would make either continuous leftward turns or rightward turns. What we found was that when the rat turned to the left more current flowed in the right electrode; when it turned to the right, more current flowed in the left electrode. In other words, the dopamine-containing terminals in each striatum released dopamine only when the animal was making turns toward the other side of the body.

It was this experiment, more than any other, that convinced me that drugs could never be the complete answer to Parkinson's disease. To do its job properly, it seemed, dopamine had to be released in a precisely timed fashion—down to a few seconds, at least. It also had to be released in a precisely localized fashion—in the left or right striatum, and most probably in small subdivisions of each striatum, depending on which movements the patient wished to make. A pill administered by mouth could never mimic this kind of precision.

The situation, then, was somewhat akin to the use of insulin in diabetes. Insulin is a lifesaver, of course, but people with diabetes generally run into long-term complications of the disease, even when they are careful about taking their insulin as prescribed. That's because the body normally regulates the release of insulin on a minute-by-minute basis to keep the blood sugar in a narrow range. Long-term complications can be reduced by frequent adjustment of doses based on tests of blood sugar. The best treatment of all is a transplant of natural insulin-producing cells of the pancreas.

How could one supply dopamine to the striatum in such a precisely regulated fashion? It could not ever be done with drugs, I thought. Something entirely different was needed.

CHAPTER 4

Curing Rats

In 1979 two studies were published—one in *Science*, the other in *Brain Research*—describing almost identical findings: both research groups had created a simple animal model of Parkinson's disease and then cured the animals, or at least alleviated their symptoms, by grafting new nerve cells into their brains.

Both studies had strong Swedish connections. One was an entirely Swedish production: the authors were Anders Björklund and Ulf Stenevi at the University of Lund, near the southern tip of the country. The other involved two Swedes—Åke Sieger and Lars Olson of the Karolinska Institute in Stockholm—as well as four Americans at the National Institute of Mental Health in Washington: Mark Perlow, Richard Wyatt, Barry Hoffer, and William Freed. Barry Hoffer was subsequently to move to the University of Colorado. (The fact that Bill Freed and I share the same last name and have similar research interests has led to confusion over the years. We are not related.)

The country of Sweden has long been home to important basic research on Parkinson's disease. The father of the field is Arvid

Carlsson. He was for a long time in the Department of Pharmacology at the University of Lund. In the late 1950s, Carlsson developed the fluorometric assay for dopamine that was used by Bertler and Rosengren to show dopamine's presence in the striatum, and he began to suspect dopamine's role in Parkinson's disease at about the same time as Oleh Hornykiewicz did. He has since moved to the University of Göteborg. In 2000, he received the Nobel Prize for an extraordinary career that included the first demonstration that dopamine was needed for normal movement in animals.

Two very well known researchers in Lund were Bengt Falck and Nils Hillarp. In the early 1960s they developed a method for visualizing dopamine and other catecholamines in tissue slices, so that the nerve cells and fibers that contained dopamine could be recognized under a fluorescence microscope. This was the method used by Dahlström and Fuxe to prove the existence of a nigro-striatal pathway, as described in the previous chapter. Falck and Hillarp's technique became enormously popular in labs around the world, so much so that their names were usually joined in the double-barreled context of the "Falck-Hillarp method."

Anders Björklund and Ulf Stenevi began transplanting dopamine cells into rats' brains in the late 1970s. In their early experiments, they found that fragments of the substantia nigra taken from the brains of fetal rats survived transplantation into the cerebral cortex of adult rats. When, months later, they killed the rats who had received the transplants and examined slices of their brains with the Falck-Hillarp method, some of the transplanted cells showed the characteristic apple-green fluorescence of dopamine, and their fibers had grown into the host rat's brain. The cells had not merely survived but had stuck to their original developmental program, which was to develop connections with the correct, predetermined targets of dopamine neurons and to produce the critical chemical transmitter dopamine. Thus, even though an adult brain has long since completed its development, it still must possess the chemical signals that tell the fibers of transplanted fetal cells where to grow.

When Björklund and Stenevi took tissue fragments from *adult* rats, on the other hand, dopamine-producing cells rarely if ever survived in the host animals. Evidently there is something about fetal cells that favors successful transplantation. We still don't know for sure what that is. Quite likely, it is the fact that the young cells have not yet put out the long fibers (axons) that connect to other brain regions. When tissue fragments are taken from adult animals the axons are inevitably cut, and this may fatally injure the cell bodies. In addition, the nerve cells in the brains of fetuses are hardier in a metabolic sense: they are much better able to survive with a low oxygen supply, for example, than are the same cells in adult animals. Other factors, such as the existence of various growth-promoting chemicals in the fetal tissue, may also play a role.

Having established that the dopamine cells survived in the host rats, Björklund and Stenevi cast about for a method to test whether the transplanted cells did anything useful. To this end, they turned to a simple animal model of Parkinson's disease that had been developed by yet another Swede, Urban Ungerstedt from the Karolinska Institute in Stockholm. You may recall that in the previous chapter I mentioned the work of Poirier and Sourkes, who destroyed part of the substantia nigra in monkeys, thus causing a Parkinson-like disorder. Ungerstedt likewise damaged the substantia nigra in rats, but he did so not with a heat lesion as done by Poirier and Sourkes, but by injecting a chemical called 6-hydroxydopamine (6-OHDA). This chemical is a specific toxin to dopamine-producing cells, so when it is injected into the substantia nigra the dopamine cells in the vicinity quickly die, and their terminals in the striatum follow suit within a few days.

Ungerstedt injected the toxin into the substantia nigra on only one side of the brain in each rat—let's say that it was the right side. Rats quickly recovered from the injections and behaved like normal animals. However, Ungerstedt showed that they weren't really normal. If he gave the rats amphetamine, a drug that causes the release of dopamine from nerve terminals, the rats would circle nonstop to the right; that is, toward the side of the damaged substantia nigra. If on the other hand he gave the rats apomorphine, a dopamine agonist (dopamine-mimicking drug), they would circle nonstop to the *left*—away from the side of the damage. Given to normal animals, neither of these drugs caused consistent turning behavior in one direction or the other.

The reason for the rats' peculiar behavior was as follows. During the days after the destruction of the substantia nigra on the right side, the dopamine terminals in the striatum on that same side degenerated, but the dopamine terminals in the left striatum remained in their normal state. Thus when the rats were given amphetamine, dopamine was released in the left striatum but not the right. As Ungerstedt found, dopamine release in the left striatum is associated with rotation of the rat to the right.

What about the *leftward* turning caused by apomorphine? Ungerstedt reasoned that during the days after the dopamine terminals in the right striatum disappeared, the nerve cells that had lost their dopamine input attempted to compensate by becoming more sensitive to the transmitter. They did this by producing more *receptors*—the molecules that sense dopamine's presence. When Ungerstedt gave the rats the dopamine-like drug apomorphine, the nerve cells in the right striatum reacted more strongly than did the normal cells in the left striatum, and thus the rats circled to the left.

Björklund and Stenevi adopted Ungerstedt's experiments to test whether their transplants had any function. They destroyed the right substantia nigra in a set of rats with 6-OHDA. Some weeks later, they transplanted a fragment of fetal substantia nigra to a location in the rats' cerebral cortex directly overlying the right (dopamine-deficient) striatum. Starting before the transplant operation, and continuing for several months afterward, they tested the effect of amphetamine on the animals.

Before the operation, and for several weeks afterward, the rats circled to the right when they were given amphetamine, in accordance with Ungerstedt's results. Starting about five weeks after the transplants, some of the rats began to show less circling behavior, as if dopamine release in the left and right striatum was becoming more balanced. By sixteen weeks, some rats were not circling at all, and indeed some were now circling in the *opposite* direction—to the left.

After several months, Björklund and Stenevi killed the rats and examined the brains with the Falck-Hillarp method to see what had happened to the transplants. The results were quite variable. In some rats, the transplants had not survived. In some, only a few dopamine cells had survived. In others, large numbers of dopamine cells had survived, and they had sent axons from the transplant into the underlying striatum, where they formed a network of dopamine terminals.

There was a striking correlation between the anatomical findings and the results of the amphetamine tests. Those animals in which there was no survival of transplanted dopamine cells continued to rotate to the right throughout the entire test period. Those animals in which transplanted dopamine cells survived and reinnervated the striatum showed a decrease in circling behavior during the test period. And the animals that showed the most abundant reinner-vation of the striatum were the ones that began to circle in the opposite, leftward direction.

The correlation between the anatomical and functional findings demonstrated clearly that the fetal dopamine cells, if they survived in the host rats, could generate a new dopamine supply in the dener-vated striatum and rectify the movement disorder. The fact that some transplants went too far, so to speak, and caused a movement disorder of the opposite kind did not arouse much attention at the time, but it probably should have raised a red flag, as we'll get to later in the book.

In 1977 and 1978, while Björklund and Stenevi were carrying out these experiments, the Stockholm–U.S. group was doing a sim-ilar study. There were three main differences. First, the transplants were pushed right through the cortex into the lateral ventricle of the brain, where they lay directly against the striatum. Thus the outgrowing fibers from the transplants had a shorter distance to grow in order to reach the striatum. Second, the researchers tested the rats with apomorphine rather than amphetamine, so the direc-tion of the animals' circling movements was the opposite of what was seen in the other study. And third, they studied the animals for only one month after transplant, rather than for several months. In spite of these differences, they also obtained strong evidence that the grafted dopamine cells had reinnervated the striatum and at least partially rectified the rats' movement disorder.

Perhaps because of the abbreviated schedule of their experiments, the Stockholm–U.S. group was the first to have a manuscript ready for publication. They sent it to *Science* in September of 1978. A few months later, Björklund and Stenevi sent their manuscript to *Science*, too, but by that time the earlier paper had been accepted, so the later one was rejected as being insufficiently novel. Björklund and Stenevi had to hurriedly resubmit their manuscript to another, less widely read journal, *Brain Research*, where it was accepted. Obliged to mention the prior study, but perhaps concerned that to do so forthrightly would risk rejection by *Brain Research* as well, they simply added a note at the bottom of the paper that stated "While this paper was under submission to *Science*, a related independent study by M. J. Perlow et al. appeared (*Science* 204 [1979] 643–647)." To an insider, this note betrayed some of the tension between the two groups: Björklund and Stenevi were willing to let the world know that their paper had been rejected by *Science*, as long as it was clear why it had been rejected—the slightly earlier submission of their rivals' manuscript.

Rivalries aside, these were two fine studies that opened people's minds, including mine, to the notion that transplantation of brain cells might really confer some clinical benefit—in Parkinson's disease and in many other neurological disorders. But Björklund himself seemed cautious on that score. In 1982, three years after his *Brain Research* paper appeared, I had the opportunity to meet him for the first time. I was traveling from Copenhagen to visit Ungerstedt in Stockholm. With Lund just across a narrow strait from Copenhagen, I called Anders and dropped by. We had just published a paper in *Nature* which showed that trained, circling rats used dopamine in their brain in direct relation to the speed and direction of their movement. It turned out his lab was just setting up our trained circling rat model for a test. We discussed his work on neurotransplants. After a long talk, I asked if he was thinking of applying the transplant to human patients. "Absolutely not," he said. "Such sensationalism would badly damage the field."

Indeed, it did seem much too radical a leap to go from these experiments on brain-damaged rats to testing the same technology on humans with Parkinson's disease. For one thing, Ungerstedt's rat "model" is really not at all like the human disease. Both conditions involve damage to the nigrostriatal pathway, to be sure, but they differ in their cause, their symptoms, and their ultimate prognosis.

What was needed to test transplants was an animal model with a larger brain and a condition closer to human Parkinson's disease. And this model should be in primates, for only primates have brains similar enough to those of humans to allow us to draw conclusions that might be relevant in a therapeutic context. But nonhuman primates, in the laboratory at least, did not naturally develop Parkinson's disease, and no one had been able to make them do so—not with drugs or surgery or any other means.

Then, out of the blue, the answer presented itself.

CHAPTER 5

The New Heroin

In the summer of 1976, a young college student by the name of Barry Kidston developed an intense interest in chemistry. Kidston lived with his parents in Bethesda, Maryland, and they gave him a chemistry set. Before long Barry was spending long hours in the basement, poring over reference manuals, setting up flasks and filters, performing chemical reactions, and purifying the resulting substances.

Unfortunately, Barry's interest in chemistry was not driven by scientific curiosity alone. During a period when his family was posted to India, he had begun to use intravenous drugs, and he now had a significant habit. He saw home production not just as a way to save money, but also as an opportunity to synthesize novel drugs—"designer drugs." He hoped to create something more potent than anything available on the street.

Barry was particularly interested in an obscure chemical called 1-methyl-4-phenyl-4-propionoxy-piperidine, or MPPP. MPPP could be synthesized from the opiate painkiller Demerol, and according to published reports it was several times more potent. Barry thought that it was likely to be in the same league as heroin, the granddaddy of the opiate family.

Showing remarkable chemical skills, Barry succeeded in making MPPP in his parents' basement. He promptly mainlined a sample and got the stupendous high he had been hoping for. Over the following months he manufactured and injected the drug repeatedly. But in November of that year he cut some corners in the synthesis of a batch. When he injected the resulting material into a vein, he felt a burning sensation in his arm. He did get high, but it was an unusual, "spacey" high. Then his life really fell apart.

Within hours, he was having difficulty moving. Within three days, he was as stiff and immobile as a statue. He could not utter a word. Finding him in this state, his parents rushed him to a hospital, where doctors diagnosed him with catatonic schizophrenia, a variety of schizophrenia that is characterized by immobility and stiffness. They treated him with anti-psychotic drugs, but to no avail. Then they tried electroconvulsive therapy. Still, he remained in his mute, locked-in state.

After several weeks of fruitless treatments one doctor, Ramon Jenkins, realized that Barry's condition resembled Parkinson's disease, so he tried giving him L-dopa. The result was little short of miraculous. Barry "woke up": he walked, talked, and acted normally. But he had not been asleep: he remembered everything that had happened to him. He had just not been able to respond to any questions or commands. So long as he had enough L-dopa in him, he was fine, but if he went too long without a dose, he reverted to his statuelike state.

Jenkins was extremely puzzled by Barry's condition. Parkinson's disease mostly affects middle-aged and elderly people: to see it in a college student was quite unusual. And Parkinson's disease is slow and subtle in its onset, but Barry had hurtled through all the stages of the disease in hours. Within three days he had entered the disease's most severe phase, a state that most patients take more than a decade to reach, if they ever do.

Jenkins knew that Barry had been using intravenous drugs of his own concoction, and he immediately suspected that something

Barry had injected was the cause of his condition. But no drug, licit or illicit, was known to have such an effect. Realizing that the problem was beyond his expertise, he referred Barry to doctors at the National Institute of Mental Health, which is also in Bethesda, not far from Barry's home.

A large number of doctors and basic researchers came to study Barry. They agreed with Jenkins that Barry had probably been put into a Parkinson-like state by some unknown drug that he had made and injected. Barry, now fully communicative, told them everything he could about the chemical reactions he had carried out, about MPPP, and how he had failed to follow the full procedure with the batch that made him sick.

The NIMH researchers were interested in the case, not just for Barry's own sake, but for other reasons, too. For one thing, other users of designer drugs might be at risk of developing the same condition. Even more significant, it seemed that Barry had unwittingly synthesized a chemical that held the key to Parkinson's disease. If this chemical could induce the disease in laboratory animals there would finally be a good animal model for the disease, with all the benefits for the development of new treatments that such a model promised. In addition, it might offer a clue to the ultimate cause of Parkinson's disease—perhaps some environmental toxin, akin to the substance that Barry had injected, triggered the slow death of dopamine cells.

One of the NIMH group was a chemist, Sanford Markey. Markey knew that MPPP was the number one suspect, but he wasn't entirely happy with that identification. After all, Barry had injected plenty of MPPP for months without suffering any unusual ill effects. Then he had suddenly fallen sick after taking some "shortcuts" in the synthesis. Would such shortcuts lead to the synthesis of a quite different and more toxic chemical?

To get a better idea of what Barry had done, Markey went to his home and examined his equipment. One hope in the back of Markey's mind was that there might still be residues of Barry's last batch remaining in his glassware. After all, Barry had not had the opportunity to carry out any further chemical syntheses in the weeks since he fell ill. Unfortunately, the young man's mother had scrupulously cleaned all the glassware.

Markey, however, was too good a chemist to be thwarted by Ajax and a Brillo pad. He saw that Barry had used a vacuum desiccator to dry the final product of the synthesis. Between the body and the lid of the desiccator was a thin layer of vacuum grease. Markey scraped up some of this grease, took it back to the lab, and used solvents to extract any organic molecules it might contain. He then analyzed the extract by mass spectrometry—the most sensitive of all analytical techniques. In this procedure the molecules in the sample are broken into fragments and then accelerated in an electric field. The fragments are sorted and identified according to their mass and their electric charge, and from the entire collection of fragments the identity of the parent molecules can be deduced. Markey found that the grease contained traces of MPPP, as he expected, as well as two other compounds, including one called methyl-phenyl-tetrahydropyridine, or MPTP.

Having identified the chemicals Barry had injected, Markey synthesized new batches of them. He injected substantial doses into rats, with the hope of causing Parkinson-like symptoms. He tried both the purified MPPP and a mixture of MPPP, MPTP, and the third chemical. The rats were affected, but only briefly. Try as he might, Markey was not able to generate a permanent "model" of Parkinson's disease.

Meanwhile, Barry himself was going downhill. The L-dopa was beginning to cause unpleasant side effects. In spite of this, Barry actually began abusing the drug—the very one that was keeping him intact. His doctors had to switch him to another, less effective medication. He also was using street drugs. About a year after his chemical misadventure, he fell into a depression. In September 1978 he overdosed on cocaine and died, right on the campus of the National Institutes of Health.

Barry's parents gave the NIMH researchers permission to conduct an autopsy. When they sliced up Barry's brain, they found that the pigmented cells in the substantia nigra—the cells of origin of the nigro-striatal pathway—had been destroyed. Within a year or less, Barry's brain had gone from that of a normal young adult to that of someone in the last stage of Parkinson's disease.

Markey and his colleagues (who included psychiatrist Glen Davis, neurologist Irv Kopin, and others) decided that even though they had not been able to fully identify the toxin that had caused Barry's condition, let alone create an animal model of Parkinson's disease, they should still publicize their results. So they wrote a short manuscript and sent it to the *New England Journal of Medicine*, the world's leading medical journal. But the *NEJM* rejected it, saying that they didn't usually publish papers based on a single case history. Then the researchers sent the manuscript to another front-rank journal, the *Journal of the American Medical Association*. But *JAMA* also rejected the manuscript, saying that there were too many authors—there were seven—for such a short report.

Finally, the researchers sent it to a new and virtually unknown journal, *Psychiatry Research*, where it was accepted. It appeared in 1979, but it attracted little attention. In the meantime, the researchers went on to other things; some of them left NIMH, and Barry Kidston and his tragic misadventure were forgotten.

So things rested until 1982. In the late spring of that year, a "new synthetic heroin" began appearing on the streets of San Jose and other towns near the southern end of San Francisco Bay. At least six people who injected this drug became desperately ill. These six included 42-year-old George Carillo and his 30-year-old girlfriend Juanita Lopez, a pair of brothers called David and Bill Silvey (aged 26 and 29), and 25-year-old Connie Sainz and her 22-year-old friend Toby Govea, a drug dealer.

Like Barry Kidston, these six individuals raced through all the stages of Parkinson's disease in hours or days, ending up in a mute, frozen state. Friends or relatives took them to emergency rooms, but the doctors who saw them were mystified. As with Barry Kidston, several of them, including the Silvey brothers, were diagnosed as suffering from catatonic schizophrenia. But how could two brothers succumb to the same rare mental illness on the same day?

Besides, drugs that combated schizophrenia failed to bring these patients out of their locked-in state.

None of the doctors who saw these six individuals had read the NIMH paper describing Barry Kidston's ordeal. Thus the matter was as baffling to them as Kidston's case had been to his doctors. The elucidation of the mystery was largely the work of Bill Langston, head of the neurology department at Santa Clara Medical Center. (Langston, with journalist Jon Palfreman, has written an absorbing book about the episode and its consequences, *The Case of the Frozen Addicts.*)

The first of these patients Langston saw was George Carillo, who had been brought to the Santa Clara Valley Medical Center from the San Jose jail, where he had been incarcerated for a parole violation—drug use. Langston learned that Carillo's girlfriend was also sick and had her brought into the hospital, too. Then a physician at the hospital, Phil Ballard, happened to visit a neurologist in Watsonville, thirty miles to the south. This neurologist, Jim Tetrud, described two similar cases he had seen—the Silvey brothers, who had been found lying motionless and unresponsive in their apartment. All four of these patients were intravenous drug users, so Ballard and Langston quickly realized that their condition was probably caused by something they injected.

Fearing that a dire epidemic was breaking out among northern California's drug users, Langston arranged for the cases to be publicized in the local media, and this led to his meeting two more affected persons, Connie Sainz and Toby Govea. Connie in particular was in terrible condition. She had lain immobile and unresponsive for weeks, she had lost a lot of weight and was seriously dehydrated, and she had developed bedsores and a crushed nerve in her leg.

Like Sandy Markey at NIMH, Bill Langston went hunting for the cause of the catastrophe. He did not have to take scrapings of vacuum grease, however. A police raid at the Silveys' home in Watsonville had already netted several samples of "heroin," and more of the drug was found in a refrigerator in George Carillo's apartment. Langston gave samples to Ian Irwin, of Stanford University's Drug Assay Laboratory. Irwin performed mass spectrometry on the samples, but the resulting pattern of fragments meant nothing to him.

Langston, meanwhile, was desperate to do something for the sick addicts. He could not help noticing that their symptoms, if they had come on slowly in an older person, would be the classic symptoms of Parkinson's disease. Thus he decided to test the effects of L-dopa on George Carillo and Juanita Lopez. The effects were just as dramatic as they had been with Barry Kidston. Within a couple of hours, George and Juanita returned to a nearly normal condition, and so did the other four patients when they were given the drug.

Although Irwin had been unable to identify the constituents of the "heroin" samples, a police lab in San Jose was more successful. That lab's mass spectrometry expert, Halle Weingarten, was able to identify the samples as containing a Demerol-like compound. This triggered in her mind a faint recollection of the NIMH paper in *Psychiatry Research*, and she told Langston and Irwin of that study. The rest was easy: Irwin realized that the mass-spectrometer patterns he had obtained most likely came from MPTP, one of the compounds identified by Markey in Kidston's vacuum grease. He ordered some authentic MPTP from the Aldrich chemical company and confirmed that the fragmentation patterns of the "new heroin" and MPTP were identical. MPTP was the likely culprit.

Langston called Sandy Markey at NIMH to discuss the similarities between the Kidston case and those in California. Langston asked about the details of the tests they had administered to rats. Had they tried giving rats pure MPTP? he asked. No, said Markey, they had only administered MPTP in a mixture with the two other Demerol analogs identified in the vacuum grease. This made Langston think that Markey had simply missed MPTP's neurotoxic effects, and he resolved to test the pure compound on rodents himself.

Meanwhile, Markey and his colleague Irv Kopin were galvanized into action by Langston's call. They still thought of the MPTP research as *their* project, even if they hadn't focused on it for three years. So they sent a colleague, neurologist Stan Burns, to California with a boldfaced mission. He was to obtain a sample of the drug Langston's patients had taken, and in addition he was to obtain a sample of Langston's *patients*. In other words, he was to recruit at least one of the six victims and take him to NIMH's hospital in Bethesda, Maryland.

This Burns accomplished, although not without first having to face a furious Langston, who accused Burns and his colleagues of scientific poaching. Since Burns had selected the two Silvey brothers, however, who were not under Langston's direct medical care, there was little that Langston could do to prevent the transfer. In the end, he also agreed to turn over a sample of the "new heroin" to the NIMH group.

Thus began a David-versus-Goliath rivalry in which NIMH, with its enormous human and financial resources, was the preordained winner. But Langston had the advantage of superior mobility, and he was able to pull a significant victory from the jaws of defeat. He, Ballard, Tetrud, and Irwin published a paper in *Science* that described their findings and identified MPTP as the likely agent that had caused the Parkinsonian syndrome. MPTP, they suggested, might be capable of generating an animal model of Parkinson's disease—something long sought by medical researchers.

The *Science* paper, which appeared in February of 1983, generated enormous publicity. It also caught the NIMH group by surprise. One thing in particular about the paper irked them. Langston could make no observations on the substantia nigra of his patients, because none of them had died. Instead, he described the substantia nigra of *Barry Kidston*. Burns had sent some microscope slices of Kidston's brain to Lysia Forno, a pathologist colleague of Langston. Langston's paper described these slices—and the destruction of the substantia nigra—without acknowledging how they had obtained the material. This was a definite breach of scientific protocol, although Langston may have felt that he acted under provocation.

By the quick publication of his findings in *Science*, Langston had drawn international attention to himself and the MPTP story. But the main scientific drama revolved around a different question: Would MPTP indeed cause Parkinson's disease in laboratory animals? Langston rushed to test the pure compound on rodents—he used some rats that the sympathetic director of a nearby research lab had given him, along with the facilities to house them. But when he injected the animals with MPTP, even in huge doses, they remained completely unaffected. Maybe Sandy Markey had been right.

Meanwhile, back at NIMH, Markey himself had been carefully repeating his earlier experiments. He now also injected the purified toxin—into rats. But here again nothing happened. One thing was now clear to both rival groups: Whatever MPTP might do to humans, it did not affect rats.

The next obvious step was to test MPTP on nonhuman primates. The NIMH and California groups tested the material on their first monkeys near the end of 1982, a few weeks before Langston's *Science* paper came out. Both saw the same effect: the monkeys came down with an instant version of Parkinson's disease. And both found that the monkeys, like the human victims, could be rescued from their immobile state with L-dopa. A robust primate model of Parkinson's disease had finally been achieved.

CHAPTER 6

Curing Monkeys

Soon after the NIMH and California groups reported on the effects of MPTP in monkeys, several groups began experiments to see if transplantation of fetal brain cells would alleviate the condition, just as it did in the rats whose substantia nigra had been damaged with the toxin 6-OHDA. One group was based at Emory University in Atlanta and was led by neurosurgeon Roy Bakay. The NIMH group also began working on transplants. A third group consisted of Gene Redmond of Yale University, John Sladek of the University of Rochester, and their coworkers. And a fourth was our group here at the University of Colorado.

Working with monkeys presents many challenges. For one thing, monkeys are costly to obtain, to house, and to work on. And we needed not just regular adult monkeys, but also pregnant females carrying fetuses of known gestational age. Maintaining colonies of breeding monkeys was very expensive. With all costs considered, a single transplanted monkey observed for a period of one year costs about $50,000. Obviously, the monkey research was never going to rival the rat work in volume. The groups who developed the basic principles of fetal dopamine cell transplantation—Björklund's and Olson's groups in Sweden, and Bill Freed's and Wyatt's groups at the NIMH—had transplanted dopamine cells into thousands of rats over the years. With these kinds of numbers, they had been able to check the importance of many variables for graft survival— the age of the donor fetuses, the site of transplantation, the number of transplanted cells, and so forth. Such a rigorous exploration would never be possible in primates.

In addition, monkeys who were given MPTP became just as ill as patients with severe Parkinson's. Just like the humans suffering with this disorder, the monkeys had difficulty eating and caring for themselves. Skin ulcers similar to bedsores could develop from insufficient movement.

Researchers at NIMH came up with a new way to administer the drug that minimized these problems. Krys Bankiewicz, a Polish-born physician and Ph.D. scientist who was doing a postdoctoral fellowship with neurosurgeon Ed Oldfield and neuroscientist Irv Kopin at NIMH, figured out a way to inject MPTP into the carotid artery on one side of the monkey's brain so that only the other side of the body would have Parkinson symptoms. Krys's model quickly became the research standard because it produced Parkinson symptoms without disabling the animal. The monkeys could still feed and care for themselves.

I firmly believe that experiments on animals, including primates, are fundamental to discovering how to treat diseases. Research using animal models of diseases has helped humans *and animals* enormously. Yet concern for animal welfare is fundamental, and I've never worked with any more animals than absolutely necessary. So, for a combination of financial and humanitarian reasons, we and the other groups kept the numbers as low as possible. Many of the published papers reported on only two or three monkeys. One of the larger studies—a five-year project recently completed here at Denver that cost hundreds of thousands of dollars— involved only fifteen animals.

Here's a "case study" of one of the early monkeys we treated with MPTP, whose rocky course illustrated both the potential

benefits and the difficulties that could be expected in applying transplantation therapy to humans. We gave this monkey five injections
of MPTP over a period of ten days, and by the end of the series he
had developed a severe Parkinson-like condition. Over subsequent
weeks, however, he began to recover, even without any transplant.
We found out later that young adult monkeys tended to recover
from MPTP treatment, whereas older monkeys do not. We still
don't know why age makes a difference in this chemical response
in monkeys. But the same is true for most neurologic diseases in
people; we simply know very little about how the human brain
changes with aging.

We had to give this monkey three more series of injections, over
a period of several months, before it developed a stable Parkinsonian state. It had difficulty walking and could no longer grasp objects
normally with either hand. Instead it kept its hands stiffly flexed at
the wrist, just as some humans with Parkinson's disease do. Because
of its difficulty with eating and moving, we were concerned that it
would not survive much longer, so we decided to go ahead with a
transplant.

As one means to reduce the number of experimental animals, we
decided to make each monkey its own control: we would leave the
left side of the brain untouched and implant fetal dopamine cells
into the striatum on the right side. Because each striatum is largely
concerned with movements of the opposite side of the body, any
improvement in the performance of the left arm and leg, compared
with the right, would likely be caused by the transplant. Thus it
would not be necessary to have a "control" monkey—a monkey
that was treated with MPTP but received no transplant.

To obtain the tissue for transplantation, we anesthetized a pregnant female monkey and removed her fetus by cesarean section. The
mother had been pregnant for about six weeks, and the fetus was
only 18 millimeters from the top of its head to its rump. Because
the groups that had worked with rats had shown that this was the
best developmental age for transplantation, we thought this embryo
would be ideal in terms of the ability of the dopamine cells to survive the transplant procedure and grow in the new host.

I dissected the fetus's brain under a microscope and cut from it the tiny region of the midbrain that contained the dopamine cells. I broke up this minute chunk of tissue into a coarse suspension of cells by passing it through a hypodermic needle. Then we anesthetized the MPTP-treated monkey and placed its head in a stereotactic headholder. This is a calibrated metal frame that allows a neurosurgeon to drive a cannula to any predetermined region of the brain.

After removing a flap of bone at the top of the head, we inserted a long hollow needle containing the cell suspension down through the cerebral cortex into the striatum on the right side. Once the tip of the needle was in the striatum, we injected a droplet of the suspension using a carefully calibrated pump. We injected a total of twenty droplets during five passes into different parts of the striatum. By these means, we hoped to cover most of the striatum on the right side. Then we sewed up the wound and let the monkey recover.

Nothing much happened for the first month after the transplant. Then, about six weeks after the operation, the monkey's left hand gradually became less stiff, and he began using it as his preferred hand to grasp the cage bars and to reach for food pellets. By three to four months after the operation he had a completely normal grasp with his left hand. He was also maintaining a more normal posture and moving about better. His right hand, by contrast, was just as stiff and immobile as it had ever been. Unquestionably, the transplant had improved the function of the monkey's right striatum.

Seven months after the operation, we decided to see if we could alleviate the monkey's remaining symptoms by transplanting fetal cells into his left striatum. We went through the same procedure as before, taking midbrain tissue from another pregnant monkey. The monkey fetus whose tissue we used was related to the one used for the earlier transplant: it came from a breeding group with the same father but a different mother. The second operation went uneventfully, and we waited expectantly for the right side of his body to improve as his left side had done.

Unfortunately, the result was quite the opposite of what we

expected. Not only was there no improvement on the monkey's right side, but within six days of the surgery its left side deteriorated to the same Parkinsonian state in which it had been prior to the first transplant. He completely lost his ability to grasp objects and his body posture deteriorated to the point that he could not keep himself upright.

Somehow, the transplantation of the second batch of cells had caused the first batch to cease functioning or even, perhaps, to die. We felt that this must involve an immunological reaction. While experiments with fetal rat cell transplants into rats had shown grafts were usually accepted without rejection, the tolerance was not absolute. Presensitizing rats with skin grafts from a different strain of rat could lead to subsequent rejection of a brain graft from that foreign rat strain. For our monkey, it seemed that we had sensitized the immune system with the first graft even though the graft itself seemed to be functioning. The second graft then "woke up" the immune system to the fact that there was foreign tissue in the body, and it then attacked both the first and the second graft. Immunologists call this phenomenon the "amnestic response."

Hoping we could reverse possible rejection, we treated the monkey with anti-rejection drugs. But after several days, these drugs had no effect, and so we made the decision to sacrifice the monkey. When we later examined his brain, we found that there were indeed plentiful transplanted dopamine cells in the right striatum—the site of the first transplant—but that the grafted tissue was infiltrated with inflammatory cells, as if it were in the process of being rejected. In the left striatum, we found no dopamine cells; apparently, the cells on this side never got a chance to take hold in their new environment before they were wiped out by the host's immune system.

This particular animal taught us a lot. First, that transplants can survive and promote improvement of function in monkeys with a Parkinson-like condition. Second, that immunological rejection seems not to be a problem with initial grafts but might be with later ones, particularly if the donors were related. And third, that unexpected things can happen—sometimes good, sometimes bad—at the cutting edge of medicine. Obviously, any human patient who volunteered for a transplant would have to be made aware of the potential risks.

In the period between 1985 and 1988, the various groups that tested the transplantation procedure presented their findings at scientific meetings. Often these presentations led to heated discussions. It seemed clear to most of us that the transplants had the capacity to improve the condition of MPTP-treated monkeys, but not all the monkeys improved, and none returned to a completely healthy condition. Some of the more skeptical researchers suggested that the improvement might be some kind of nonspecific effect of the transplants—that simply sticking needles into the striatum might benefit the monkeys, rather in the same way as acupuncture is said to cure a variety of ills.

Even more contentious was the question of what the results meant for clinical medicine. Should we go ahead and try the methodology on people with Parkinson's disease? At one meeting organized by Anders Björklund and held in New York in 1986, quite a dispute broke out on this subject after the conclusion of the presentations. Constantin Sotelo, a leading French neuroscientist, asked for human trials to be put off for several years at least, until the cause of Parkinson's disease was better understood. After all, he said, whatever mysterious process kills off the patient's own dopamine cells might kill off the transplanted cells, too. In that respect, he said, the MPTP-treated monkeys are not a good model for the human disease, because MPTP, once it has done its damage, leaves the body. Thus it has no chance to attack cells that are transplanted at some later time.

Bill Freed's colleague Richard Wyatt took issue with this. He pointed out that even if the patient's own disease process does attack the transplanted cells, it should not do so any faster than it kills the patient's own cells—which meant slowly. "Once we have the technique worked out in monkeys," he said, "neurosurgeons should be able to apply it in humans. If you could buy a patient four years of life it would be worth it. If he gets ten years, that's magnificent."

George Pappas, of the University of Illinois, was critical of the basic notion of transplantation. "Transplantation is not the answer for anything," he said. "Just as artificial lungs were not the answer to the problem of polio, transplants are not going to be the answer in the long run. Doing transplants will simply delay finding out what goes wrong in Parkinson's disease. We must understand the difference between technology and basic research."

Efrain Azmitia, of New York University, brought up an issue that I also had been thinking a lot about. He was concerned that the effectiveness of the transplant surgery would not be analyzed in a sufficiently rigorous way. "Maybe we should go so far as to have controls in the human trials—patients who get operated on but actually receive no grafts. It sounds cruel, but I think these are the kinds of things we have to do."

Olle Lindvall, of the Lund group, objected to that idea. "I think at the moment it would be unethical to do any sham procedures," he said. "If there were spectacular effects, then we should consider doing sham surgery."

Azmitia pointed out that placebos are regularly used in testing of drugs, so why shouldn't they be part of surgical trials, too? Olson didn't buy it. "Imagine that the procedures worked very well, except in two patients, and afterward we had to say to them, 'Sorry, but you were the controls.' That would be a very tough situation." Azmitia wasn't satisfied. "Imagine those two patients improved as much as the others," he said. "Then it would be just the effect of implanting the cannula in the striatum, not of the transplanted cells."

Stanley Fahn, of Columbia University, weighed in strongly on Azmitia's side. He mentioned various useless operations that were thought to be effective for a long time because no one did a controlled trial. "It always amazes me how careful basic scientists are to have controls in animal experiments," he said, "but once they start working on patients the controls disappear out the window and we're left with bad treatments for years. The history of medicine is replete with accounts of leech treatments and the like that were used before scientists or doctors began using controlled approaches to treatment and testing. We must be even more careful because of the seriousness of this surgical procedure and the potential harm that can be done to some of these patients." Several years later, as we'll see, Fahn became a close colleague of mine, and the line of thinking he expressed at the meeting greatly influenced how we designed our experiments.

John Sladek, of the Yale-Rochester group, summarized the basic dilemma. "For all we know, the grafts may grow out of control and cause some kind of obstruction and kill the patient," he said. "But on the other hand, when I recall so many unhappy patients depressed at facing a progressive disease that will probably kill them in ten years, I don't want to see us stand still either."

The ethical question of using tissue from aborted fetuses was also discussed at the meeting. All the participants who spoke on this matter felt that the use of this tissue was ethically acceptable, provided that certain precautions were taken. In particular, the decision to have the abortion should be completely separate from the decision to donate the fetal tissue for transplantation.

Yet at least one participant at the meeting, neurosurgeon Erik-Olof Backlund of the Karolinska Institute, did have serious moral reservations about the use of fetal tissue. These reservations were based on a strong religious belief that abortion was wrong, and that taking advantage of an abortion to treat a patient was therefore wrong, too. Backlund did not attempt to promote his views at the meeting. In fact, his anti-abortion stance had motivated him to try transplanting a quite different kind of dopamine cell—one that was not derived from fetuses at all. He used tissue taken from the adrenal glands.

The adrenal glands sit like little caps on top of the kidneys. They contain two very different kinds of glandular tissue. The outer rind or *cortex* of the glands secretes steroid hormones, while the inner portion, known as the *medulla*, secretes catecholamines—mainly adrenaline and noradrenaline. The cells of the adrenal medulla also synthesize dopamine, but only for their own use—as an intermediary in the production of the other two hormones. They do not usually secrete dopamine in significant amounts.

It turns out, however, that the glandular cells of the adrenal medulla can be induced to become somewhat more like neurons in certain conditions. Treatment with a chemical called nerve growth factor will make the cells assume a neuron shape and even send out fiber connections like other neurons do. Because the cerebrospinal fluid contains some nerve growth factor, transplants placed in the striatum, where they are exposed to cerebrospinal fluid in the brain's ventricles, could survive and send out fiber process into the brain.

Bill Freed and his colleagues, as well as other groups, tested whether transplanted adrenal medulla cells would correct the turning behavior of 6-OHDA–lesioned rats, in the same way that transplanted fetal dopamine cells did. The researchers did in fact get fairly positive results with the adrenal cells, and this motivated the Stockholm group, led by Backlund, to test the procedure on two patients with Parkinson's disease. These operations took place in 1982. Not only were Backlund and his colleagues able to sidestep the use of tissue from aborted fetuses, they also got away from problems of immune rejection, because the adrenal cells could be harvested from the very patient who was to receive them.

Backlund's initial two patients did not experience any lasting benefit from the transplants, nor did two more patients on whom the Swedes operated in 1985. But meanwhile a Mexican neurosurgeon, Ignazio Madrazo, had picked up the ball and was running with it. He claimed dramatic improvements in his patients, and when his report appeared in the *New England Journal of Medicine* in the summer of 1987, it ignited a wave of enthusiasm for the procedure. Soon neurosurgeons were performing hundreds of adrenal medulla transplants on patients with Parkinson's disease around the world, including in the United States.

Adrenal medulla transplants don't work—that much is now crystal clear. The transplanted cells may survive for a while, and they may even provide some temporary relief of symptoms, but within months the cells die, leaving the patient in as bad or worse a condition than before. The effect of the adrenal medulla "craze" of the late 1980s was mainly to sour researchers and the public on the whole notion of transplantation as a possible therapy for Parkinson's disease.

In any event, the discussions at Björklund's meeting and the other research meetings during the mid-1980s made it obvious that researchers were seriously considering, perhaps even planning for, the transplantation of fetal dopamine cells into humans with Parkinson's disease. No one said out loud that they were intending to do it, but it was in the air. Even Anders Björklund, who had been so negative about the idea when I had asked him in 1982, and who had not been doing any experiments with MPTP, expressed himself more positively when I asked him about it again during a visit to Lund in 1984. "If it works, people will accept it," was his laconic response. To me, that meant that Björklund was planning to go ahead.

Given the air of expectancy, it was good that all these technical and ethical issues got heard. Yet I couldn't help wondering what would happen if such a group of experts as were present at the New York meeting were made into a committee to decide whether to go ahead with the transplants. Almost certainly there would be endless discussions and no decisions. Someone simply had to take the bull by the horns.

Back in Denver, I thought carefully about how to proceed. In 1986, as we were getting our first favorable results with the MPTP-treated monkeys, I decided to get together a group of clinicians and researchers who might be participants in a clinical study of fetal-cell transplantation. The idea was that this group would collectively look at the progression of our animal experiments and would let us know when and how it would be safe to proceed.

One of the people in the group was Stuart Schneck, professor of neurology at the University of Colorado. You may recall from chapter 1 that he was the person who first correctly diagnosed Don Nelson's condition. Stuart had enormous experience with Parkinson's disease, having been a participant in the first multicenter trial of L-dopa back in the 1960s, as I mentioned earlier. Of course, I had seen many patients with Parkinson's disease myself, but Stuart had followed patients over *decades*. He was intimately familiar with all the vagaries of the disease—the inexorable worsening, the paradoxical remissions and improvements, the spectacular benefits and the heartbreaking disappointments of drug therapy, as well as terrible psychological trauma suffered by patients.

There was a particular historical factor that made Stuart sympathetic to the notion of a transplantation study. He had collaborated with Tom Starzl, the surgeon who pioneered liver transplantation while at the University of Colorado in the 1960s. Stuart's role had been to look for evidence of brain damage in autopsies of liver-transplant recipients who died. And they did die: of the first seventeen people who received a liver transplant—all of them at death's door when they came to surgery—not one survived. Starzl showed incredible strength of character—some would say recklessness—to persist in the face of these daunting results. But each patient taught him something, and eventually he was able to improve his techniques to the point that liver transplantation became a routine, lifesaving operation. Only the shortage of donor livers now limits the application of the treatment.

Stuart was so impressed by Starzl's work that, when I began mentioning transplantation as a possible therapy for Parkinson's disease, he didn't reject the idea out of hand as many neurologists would have done. Instead, he joined our planning group. He began thinking about which patients in his extensive practice might make appropriate candidates to receive transplants. And he made a crucial contact that allowed our project to take off, as we'll see later.

Among the other people in the planning group were psychiatrist Martin Reite and postdoctoral fellow Jerry Richards, who were collaborating with me in the MPTP work in monkeys, and three neurosurgeons, Glenn Kindt, who was head of the neurosurgery division, Bruce Tramner, and Andrew Rhea, who would likely be the ones to actually perform the first operations, if it got to that point. In fact, both Bruce and Andrew left the university shortly hereafter. To our good fortune, in 1987 Dr. Kindt hired an extraordinary neurosurgeon, Robert Breeze, direct from his residency at the University of Southern California. Bob's willingness to solve a new problem in neurosurgery made our transplant program possible. A radiologist, Kathy Davis, also joined the group; her role would be to conduct magnetic resonance (MRI) scans of the patients prior to surgery, in order to determine exactly where their striatum was located.

Another person we recruited was Trent Wells. Trent, who is now in his eighties, is quite a character. He was a fighter pilot in World War II and saw action over Europe. After the war, he revealed an inventive genius; he developed new kinds of equipment for stereotactic surgery and founded a company that became the leading manufacturer of such equipment. Later, when our program was struck by disaster, his inventiveness helped get us back into business.

Two immunologists, Henry Claman and David Talmadge, also joined the group, because we had to think about the possibility that the transplants would be rejected by the patients' immune systems. Also involved was David Rottenberg from Cornell, who had been a medical school classmate of mine at Harvard. He was director of Cornell's PET(positron-emission-tomography)-scan facility and his scans would be key to determining whether our transplants survived. Another participant was John Barrett, from the University of Miami, who had developed methods to keep neurons alive in cold storage.

The planning group was one-sided in the sense that it consisted only of people who thought fetal-cell transplantation was a form a therapy worth pursuing, and who had no objections to the use of tissue from aborted fetuses. I knew that if the treatment ever came close to being a reality, other independent committees would have to review the ethical and other issues. Our group was a collection of people with the skills and motivation to turn a theoretical possibility into a clinical reality. Over the ensuing two years, we met repeatedly to discuss the progress of our monkey experiments and to hammer out all the technical details. We gradually approached the point where we felt ready to proceed. But meanwhile, events elsewhere were making it harder and harder for us to do so.

CHAPTER 7

Politics

In December of 1987 I got word over the academic grapevine that Anders Björklund, Olle Lindvall, and their Lund colleagues had performed the first two transplants of fetal dopamine cells into patients with Parkinson's disease. The surgeon was Stig Rehncrona: Backlund had withdrawn from the team a year earlier, in part because of his reservations about the use of fetal tissue.

A part of me was disappointed that we would not, after all, have the opportunity to carry out the first fetal-cell transplant. Yet I was also pleased that the Lund team had done the operations, for if anyone could be relied upon to do them in a careful and competent manner, it was this team. As I later learned, Björklund and Lindvall had assembled a team of experts to plan and direct the study, just as we had done, and they had meticulously rehearsed every step of the procedure.

I also knew that they were unlikely to make premature or exaggerated claims for success. In fact, the Lund group did not speak publicly about their work for over a year after the first two operations were performed. During this time, they followed their two patients—both women about fifty years of age—in minute detail, documenting every aspect of their illness on an almost daily basis. All I knew, though, was that the operations had been done and that no disaster had struck.

In January 1988 our planning group submitted a request to the University's Institutional Review Board, or IRB, for permission to carry out our first transplants. IRBs are watchdog committees that review applications to conduct clinical trials. Their main concern is to look out for the interests of the patients who volunteer for the trials; they check that the patients will not be exposed to unnecessary risks and that they are properly informed of whatever risks may be unavoidable. The IRBs also examine the proposals to see if the trial really is likely to obtain the hoped-for information, and how useful that information will be. Only if the benefits of acquiring that information outweigh the risks to the patients will the IRB approve the application.

Perhaps because the University of Colorado has such a tradition of pioneering research in transplantation, the IRB looked favorably on the general idea behind the study and did not ask us to make many changes. I was relieved to get the IRB approval so quickly—I felt that we could now proceed rapidly toward our first transplant operation. But unknown to me, a major impediment to our project was building.

Late in 1987, about when the Lund group were doing their first operations, a group at the National Institute of Neurological Disorders and Stroke (NINDS) also decided to go ahead with fetal-cell transplants. This group was led by Irv Kopin, director of intramural research at NINDS, who had been involved in the study of MPTP victim Barry Kidston, and neurosurgeon Ed Oldfield. Oldfield and Kopin's proposal was approved by the institutional review board at NINDS, just as ours had been at the University of Colorado. But that was not the end of the approval chain. Approval was next sought from the head of the National Institutes of Health, James Wyngaarden.

Rather than okaying or rejecting Kopin's request, Wyngaarden sent it further up the political ladder to Robert Windom, Assistant

Under-Secretary of Health in the Reagan administration. Windom probably checked with *his* superior, Secretary of Health and Human Services Louis Sullivan, and Sullivan probably checked with the White House. Whatever the exact line of communication, Windom finally issued a moratorium on federal funding, not just for Kopin's project but for *all* studies involving transplantation of human fetal tissue into humans. And he ordered Wyngaarden to convene a high-level panel to review the medical, legal, and moral issues raised by this kind of research. Depending on the conclusions of this committee, the moratorium might be lifted.

I must say, when I heard about these developments, I was disgusted. The Reagan administration had decided to put its political spin on research. That attitude persisted at NIH through the Bush years under then-director Bernadine Healy. It is my understanding that morale suffered and that many outstanding scientists left the NIH during those years.

Wyngarden did arrange a review panel for the fall of 1988. Even before it convened, however, someone in the administration leaked to the press a draft of a ban on federal funding for *all* research on tissue from aborted fetuses—whether it was to be used for transplantation or any other purpose. The draft had been written by Gary Bauer, a conservative advisor to President Reagan. (In the 2000 primary elections, Bauer ran for the Republican presidential nomination, largely on an anti-abortion platform. He received less than 1 percent of the vote in the primaries.) The leaking of this draft right before the initial meeting of the review panel seemed like a broad hint that, whatever the panel recommended, the administration had already made up its mind.

The review panel met for the first time on September 14, 1988. Of the twenty-one panelists, at least two were sure to vote against fetal-tissue research whatever the testimony. One of these was Father James Burtchaell, a theology professor at Notre Dame University. To Burtchaell, abortion was murder and anyone who used fetal tissue that resulted from abortion was an accessory to murder. It was that simple. Conversely, there were probably one or two scientists on the panel who were certain to vote in support of fetal-tissue research.

For most of the panelists, though, it seemed worth discussing practical questions like: Would the fact that fetal tissue was being used for transplantation cause the numbers of abortions to increase? Conceivably, this might happen because the researchers would put pressure on women who were considering abortion to undergo the procedure, or would provide them with some kind of inducement to do so. Or there might be a more nebulous connection: federal approval of fetal-tissue research might be seen as an official validation of abortion.

Witnesses were divided on these issues. Lars Olson, the Karolinska Institute scientist, testified that the medical use of fetal tissue in Sweden had had no influence on the rate of abortion in that country. Witnesses from anti-abortion groups were equally emphatic that such use *would* encourage abortion. What is more, conservative ethicists argued that women who had abortions had no moral right to sign consent forms for use of the fetal tissue, because they had abdicated their status as guardians of their future children by consenting to the abortion.

Another question that came up was whether researchers could use tissue from *spontaneous* abortions. Such use seemed to be more acceptable to the anti-abortion witnesses, because there was no intent to harm the fetus. But the scientific witnesses argued that tissue from spontaneous abortion was dangerously abnormal: spontaneous abortion, which is very common, usually occurs in response to maldevelopment of the fetus, and the tissue from such fetuses is likely to be infected and to have chromosomal abnormalities.

After meeting for three days, the committee adjourned for a few weeks. But I was increasingly convinced that regardless of what the committee recommended the ban on federal funding was likely to be continued by the Reagan administration or by its successor. (Reagan's vice president, George Bush, was seeming like a probable winner in the upcoming presidential election.)

It was clear that all the discussions had to do with federal funding of human transplant operations. Ironically, it was considered perfectly acceptable to use federal funds to acquire human fetal tissue for laboratory research, including implanting human fetal cells into experimental animals. The only issue was the use of those same cells to treat suffering humans, including those with Parkinson's disease.

Our institutional review board inquired about whether we were planning to use federal funds. We indicated we would be using private funds. Because our first patient, Don Nelson, had been admitted for a few days of evaluation to the University's Clinical Research Center, which received NIH funding, we refunded money to the Center, lest there be any misunderstanding later.

But if federal funds were not going to be available, where else could we get money to do these procedures, which we estimated would cost at least $30,000 each? Would we have to ask the patients themselves to pay for them? That was an unappetizing prospect, especially for the earliest patients, to whom we could give the least assurance of any benefit from the procedure.

Stuart Schneck got us out of this dilemma. Stuart had an elderly patient with Parkinson's disease by the name of Charles Edwin ("Ed") Stanton. In 1956 Stanton, an architect, had married May Bonfils. May was one of two sisters who inherited considerable fortunes from their father Frederick Bonfils, longtime publisher of the *Denver Post*. At the time of their marriage Ed Stanton was in his forties, and May Bonfils was in her seventies.

The Bonfils daughters were energized philanthropists. May's sister, Helen, was devoted to the theater and poured her money into Broadway productions. May, on the other hand, was a devout Catholic who gave money to found a Franciscan monastery in Denver. In addition, May wanted a traveling companion and someone to manage her money—roles that Stanton fulfilled very well. Thus, if their marriage was an "arrangement," it was a happy one for both of them.

May and Ed lived at Belmar, a million-dollar, marble-clad mansion that was an exact replica of Le Petit Trianon in France, set in a 750-acre estate west of Denver. Their bed was once owned by Marie Antoinette, and some of their furniture had belonged to Queen Victoria. "The twentieth century does not exist" was one of May's favorite remarks.

Ed Stanton did not spend all his time at Belmar—he maintained apartments in San Francisco and Paris, and he liked to go on lengthy cruises—but he did manage May's finances with great success, so that by the time she died in 1962, her fortune was many times larger than it had been when they married. Stanton moved into an "apartment"—an entire floor of a high-rise building, which he filled with Greek statuary and Renaissance art. Although Stuart Schneck was not in the habit of making house calls, he did so in Stanton's case, and in fact the two men became good friends. But by the mid-1980s, when Stanton was in his seventies, his condition had deteriorated badly, and his older brother Robert had taken over his financial affairs.

In the fall of 1986, I applied to the Bonfils-Stanton Foundation for funds to help finance our transplant research in monkeys. At year's end, on December 29, eighty-three-year-old Robert Stanton called me to say he would like to discuss our funding request. I said, "How about nine A.M. tomorrow?" He came to visit the lab the next morning. We presented our research, visited the animals, and then he wrote me a check for $150,000 from his and his brother's personal funds, not from their $250 million foundation. This extraordinary gift gave us the financial base for our monkey studies and paved the way for the first human transplants.

Unfortunately, Ed Stanton's condition continued to worsen, and he died in 1987 before we did the first transplant. The brothers' generosity, combined with gifts from the National Parkinson Foundation and from our instrument builder, Trent Wells, made it possible to go forward without federal funds.

As I mentioned at the beginning of the book, Don Nelson noticed the first symptoms of Parkinson's disease in 1969, but the disease wasn't diagnosed correctly until four years later, when he was seen by Stuart Schneck here at the University of Colorado Medical Center. Don's experience with incorrect diagnoses is far from unique; many of my patients have gone through an equally frustrating period before learning the real identity of their ailment. The initial signs and symptoms of Parkinson's disease may be quite subtle and could be due to a number of disorders such as the supposed ulnar nerve problem in Don's case. Also, doctors may be aware that Parkinson's disease is on the list of possible culprits, but they may not want to burden the patient with such a depressing diagnosis while some other cause is still in the cards.

A special factor in Don's case was his age. Parkinson's disease usually strikes after the age of fifty. When Don told the neurosurgeon who had referred him to Stuart Schneck what Stuart's diagnosis had been, he told Don: "That's impossible—you're too young." But Parkinson's disease does strike some people in their thirties or even younger.

Don was lucky to end up in the care of Stuart Schneck, who at that time was already an expert on the disease and its treatment. Stuart began Don on gradually increasing doses of L-dopa, along with some other drugs. This was before the introduction of the L-dopa/carbidopa combination (Sinemet), so Don had to take high doses of the pure L-dopa, which provoked extreme nausea, but it was worth it: he got back pretty close to a normal state.

For a few years, things went well. But gradually the right side of Don's body also became affected by the disease, though never to the same extent as the left side. Don also developed increasingly florid dyskinesias. By 1973 he was finding his managerial job too stressful, and he had to quit. After that, he and his wife, Carolyn, ran a fast-food restaurant. This also eventually became too burdensome, and the Nelsons quit the business in 1980, when Don was forty-three. He tried to find other work, but his dyskinesias were a major impediment. "Interviews that were once a smooth situation were now very tense moments," says Don. "I could not control my arms from shaking and this, added to my age, was negative, and employers politely told me I was overqualified."

Parkinson's disease brings out the best or the worst in people; in the Nelsons' case it brought out the best. Even after he had to stop working, Don remained physically and intellectually active, and very focused on finding ways to alleviate his condition. Carolyn, like many spouses of Parkinson's disease patients, had to carry a tremendous burden—as caregiver to her husband and now as breadwinner. Luckily, she is an energetic, optimistic woman. To make money, she took over the tiny bridal wear department at the local J. C. Penney's store, and she developed it to the point that her outlet was the chain's nationwide leader in sales.

By 1982 Don was beginning to have trouble walking. One moment he might be walking fine, and the next he would stumble and fall forward. He took to using a cane. Increasingly, walking lost its automatic quality; Don had to "will" each step, otherwise his feet would simply stick to the floor. This, and the dyskinesias, made it increasingly embarrassing for Don to leave the house. He began to dread every outing, whether it was to the store, church, or to any other kind of interaction with the public.

Other problems plagued him as well. Even though he wrote with his less-affected right hand, his handwriting deteriorated to the point that it was illegible, even to himself. His voice lost its strength and clarity. He also became chronically constipated—a common though poorly understood complication of the disease.

Stuart Schneck did the best he could to alleviate Don's condition by carefully adjusting his drugs and by adding new ones. But it was a losing battle. By 1986, Don was noticing some kind of deterioration on an almost monthly basis. He was desperate to try something new. To raise his spirits, Stuart told him that we were working on a possible new treatment, but he didn't spell out what it was.

Stuart had mentioned Don to me as a possible candidate for our first fetal-tissue transplant. Late in 1987, I got a letter from Don himself expressing his interest in learning about our research. So I invited Don and Carolyn to visit, which they did. I showed them videotapes of our MPTP-treated monkeys, before and after we gave them transplants. Don was moved by the sight of the affected monkeys: he recognized instantly that they were enduring the exact same disorder as he was.

There were a number of reasons why Don seemed like an excellent candidate to be our first transplant recipient. First, he was seriously debilitated by the disease and drugs were not keeping it in check. An experimental, potentially risky therapy is not something one wants to test on people who would have a reasonable quality of life without it. Second, the fact that Don's disease primarily affected his left side was an advantage. We had decided that it would be prudent to transplant cells into only one side of the brain in our first patient. With Don, we could plan to do the transplant into the right hemisphere, and then watch to see if there was any improvement in the left side of his body. And thirdly, Don did not have any other major medical problems: he seemed like someone who, if only his Parkinson's disease could be improved, would return to a fully normal and active life.

Besides these medical issues, there were psychological factors that made Don a good candidate. Don was strongly motivated to volunteer for the new treatment. He is also an intelligent, observant man with an accountant's interest in the details of his disease; he had kept a record of his condition over a period of many years. That trait was invaluable to us, because we were going to ask Don to monitor his own condition in great detail, both before and after the surgery.

Finally, what made Don a good candidate was his family. Not only Carolyn, whom I soon found to be a tower of strength, but his children and other relatives were all close—physically and emotionally. That level of support can make the difference between success and failure.

After a few days of further discussions among our planning group and with the Nelsons, we collectively decided that Don Nelson should be our first patient to receive a transplant. This was in February of 1988. We went through all the risks that the transplant carried with it: the possibility of a stroke or infection from the surgery, and the possibility that the transplant might worsen Don's disease rather than alleviate it, or that it might have quite unexpected and harmful side effects. These risks were also spelled out in the consent form Don signed.

One issue was particularly difficult to resolve: should we treat Don with immunosuppressant drugs? In general, humans reject tissue grafts from unrelated donors. Whether the transplanted tissue be kidney, heart, liver, or skin, the body recognizes the grafted tissue as foreign and launches a massive attack that destroys the graft within a week or two. Thus, transplant recipients are usually treated with large doses of immunosuppressive steroids as well as another powerful drug, cyclosporin A. Often, these drugs allow the transplant to survive indefinitely, but they do so at a heavy cost, both financially (upward of $10,000 per year for drugs and associated medical expenses) and medically (side effects can include hypertension, kidney impairment, infections, and even cancer).

There were good reasons to think that transplants of human dopamine cells to the brain would be less prone to immunological rejection than transplants elsewhere. Because the blood vessels passing through the brain are much less "leaky" than those in the remainder of the body, the brain is largely beyond the reach of the body's immune system. Most important, transplants of fetal brain cells in rats and monkeys had survived without immunosuppression. The protection from the immune system is not absolute. Transplants of tissue from a foreign species, such as pig cells into a rat's brain, are aggressively rejected. Animals can also be sensitized with skin grafts, and they will then reject subsequent transplants into the brain. That, at least, was the general experience of researchers such as ourselves who had carried out transplants in laboratory animals.

Would brain transplants in humans be rejected without immunosuppression? We couldn't be sure. I asked David Talmadge, the University of Colorado immunologist who came up with the idea that the diverse antibodies in our blood are selected from an almost infinite number of possibilities by a clever cloning system intrinsic to our immune system. I also asked Henry Claman, who was first to realize that the lymphocytes in our bodies originate in two different locations, the thymus gland and the bone marrow, and that these two types of immune cells have different functions. I checked with a third immunologist, Kevin Lafferty, who was pursuing fetal pancreatic transplants to treat diabetes. Because diabetes is an autoimmune disease in many patients, the challenge of his research was great. I also called Richard Hodes and Ron Schwartz, old friends from medical school who were now at NIH. Richard is now director of the National Institute on Aging. Ron has made important contributions on how receptors on our lymphocytes work.

All these consultations left me a bit perplexed, because none of these experts was sure if immunosuppressants would make a difference. In the end, I decided that I would rather have the first transplant fail than commit future transplant patients to unnecessary immunosuppression. Once immunosuppressants were used, it would be difficult to discontinue their use in the future. We decided not to immunosuppress our first patient, but to do so with some of the subsequent patients.

In May, we sent Don to the Memorial Sloan-Kettering Hospital in New York for a PET scan. This procedure—its full name is *positron-emission tomography*—involves injecting a radioactive version of L-dopa (18-F fluorodopa) into the bloodstream, and then imaging the distribution of radioactivity in the patient's brain with a special-purpose scanner (we'll have more to say about this technique later in the book). Healthy terminals of dopamine cells will take up the radioactive tracer as if it were L-dopa, so the striatum normally "lights up" with this procedure. In Don's case, however, the uptake was abnormally low, especially on the right side. This result was consistent with the advanced state of Don's disease and the worse condition of the left side of his body compared with the right. We hoped that, by comparing Don's PET scan before and after surgery, we could get objective evidence as to whether the grafted cells survived and functioned in his brain.

We also needed to monitor Don's condition in as thorough a way as possible. With Parkinson's disease patients, a single examination in the doctor's office can give an inaccurate impression because the symptoms are so variable over the course of the day and with varying levels of medication. It was essential that we had a true picture of Don's baseline condition prior to the operation. So we admitted him to the University Hospital for extensive neurological tests. And then we began Don on a program of self-testing at home.

For these tests, I set up some electronics to collect data in an automated fashion. One test measured the speed of Don's finger movements. I put together a box with some buttons on it, controlled by a Commodore 64 computer—now a museum piece, but then an excellent computer: cheap, easy to program, and easy to connect to homemade equipment. A cue would appear on the computer screen, and Don had to press the appropriate button as fast as possible. Or he had to alternate rapidly between two buttons. I also set up a simple apparatus to measure Don's walking speed: I purchased some infra-red photocells from Radio Shack, laid them out on the floor in Don's home, and connected them to the same computer. Don would have to walk as fast as he could, and the photocells would record his walking speed. Don had to do these tests several times a day, beginning early in the morning before he had taken his first dose of Sinemet.

Don cooperated magnificently with these tests. He demonstrated the same precision he used to document his illness. He not only did them all at the appropriate times for months on end, he also recorded the slightest glitch. So if there was an odd result one day—if the computer showed Don moving at Olympic speed, for example— I would see a note from Don saying "Cat jumped in front of photocell" or the like.

Bob Breeze and I met to discuss his plans for the operation. We intended to make many penetrations of the brain in order to put each deposit of fetal tissue close enough to the others to bridge and fill the entire striatum on the right side of Don's brain. Studies by Olson's group in Stockholm and Bjorklund's group in Lund had shown that human dopamine neurons could span at least 4 millimeters in a rat's brain. Basing our plans on the results of these experiments, we decided to make a series of needle passes from the top of the head at 4-millimeter intervals. Bob calculated that that would mean a total of ten needle passes into the brain.

Running needles into the brain is nothing to undertake lightly. It is done fairly commonly, usually with the purpose of obtaining a biopsy from a suspected brain tumor. But according to our estimates, a patient who undergoes a single needle pass into the brain has about a 1 in 500 chance of experiencing a catastrophic hemorrhage. The math was simple: making ten needle passes into a patient's brain would increase the likelihood of a hemorrhage to 1 in 50. We could only hope that Don would not be that one.

This testing and planning continued through the summer of 1988. By the fall, we had our patient ready, we had approval from the IRB, and we had funds from Stanton. We had worked out detailed plans for the operation. All we needed was the fetal tissue.

Starting in 1986, I had been working with several Denver-area abortion clinics to see if any recognizable tissue could be recovered from early embryos, when the whole embryo is only about one inch long. Because abortions destroy the embryo, usually nothing is recognizable. It took more than a year before we could recognize fragments of tissue that contained the critical dopamine neurons. Even with that experience, tissue was hard to recognize and recover. At least 90 percent of the time, nothing was recognizable.

For Don and Carolyn, though, this waiting period was especially difficult. On four occasions at least I warned Don to be ready for surgery—which meant not eating anything all day and not taking his usual medications—only to find that we had no usable tissue to transplant, and we had to let Don and his family off the hook.

Another reason that I found this waiting period so vexing was my concern that news of our plans might leak out. Although what we were doing was completely aboveboard, legal, and approved by all the necessary committees, I knew there were people both inside and outside the medical research community who might try to put a stop to our project if they got wind of it. The Reagan administration's moratorium applied only to the use of federal funds for this kind of research, and we were not using federal funds. But some people were interpreting the moratorium as a de facto ban on all fetal transplant work, regardless of how it was financed. Thus, there was only one way to avoid the inevitable confrontation and the prospect of further delays, and that was to do the operation.

Our visit to the clinic on November 8 started out like all the other days. My assistant Shar Waldrop loaded sterile bottles, pipettes, petri dishes, and all our other paraphernalia onto a little cart—actually a small red wagon—which was the right size to do the job and still fit into a car. We took the wagon and drove the few miles from my lab to the clinic. We entered the clinic through the back door, with Shar pulling the wagon behind her. In a small room that the clinic had set aside for our use, we set up a sterile area and laid out our glassware and instruments and a dissecting microscope. We donned sterile gowns, gloves, and masks, and waited.

After thirty minutes or so, we got the word: the first abortion of the day had been completed. Like most abortions, this one had been conducted during the first trimester when the embryo is very small, about one inch long. A clinic staff member came into our room with a glass bottle containing the tissue fragments and fluid from the abortion. Under the microscope, we looked for a piece of tissue that contained the dopamine cells in the midbrain of the embryo, a fragment about one-eighth of an inch in size. Shar took a brief look and then yielded her place to me. When I saw the pieces under high magnification it was clear that, once again, the tissue was unrecognizable and, therefore, of no use. Shar put away the used glassware, petri dishes, and instruments and set up new, sterile ones. Then we waited again.

These waits have often been an occasion for me to ponder about abortion and about the women who come to these clinics. The decision is a difficult one for any woman to make. About 70 percent of Americans agree that women should be allowed to make this decision for themselves, while about 30 percent do not want them to have the option. The right to choose abortion has been guaranteed by the Supreme Court. When a woman considers abortion, she has to weigh many factors—her life and relationships and future plans, and her own sense of what is right or wrong. I went through college and medical school at a time when abortion was illegal in many states. I saw many young women make tragic decisions to end an unwanted pregnancy. Often, the woman would travel to a state that allowed abortions, if she could afford it. Others made the riskier choice of secret, illegal abortions done by people with uncertain credentials in ill-equipped settings. Of course, some couples married. If the relationships were strong, marriages succeeded. If pregnancy was the primary bond, marriages ended quickly and bitterly. A child was left in the middle.

Women choosing to end their pregnancies come to clinics with these complicated thoughts. They are counseled about their options. The grimmest choice—a back alley abortion—has been eliminated, but all the other issues about relationships and reproductive plans remain. Women who have abortions have decided that it is the best choice for them.

What happens to tissue after abortion? It is usually discarded. Tissue donation is an alternative. If we find a fragment that contains dopamine nerve cells, we ask the woman if she would be willing to donate this tissue to help our research and to treat patients with Parkinson's disease. If she says yes, she signs a consent form, and we keep the tissue. If she says no, tissue is discarded, as it usually is. Laws prohibit money being offered for tissue, nor may a donor direct that tissue be used to treat a specific person. The alternatives are simple: discard the tissue, or use it for research.

A little while later another procedure was completed, and we went through the same ritual, with the same result. Then another, and another. Between times I read papers, made phone calls, or chatted with Shar. Hours went by, and the frustration was getting to me. Were the fates conspiring to impede our plans, just when everything was in place, the operating team was ready, and one brave patient was fasting—as he had fasted so many times already—in hopes of surgery that day?

I just had to take a break. Then it occurred to me that I had a perfect excuse to take one. It was Election Day, after all, and I of all people had reason to vote.

Leaving Shar in charge, I hopped in my car and drove to my polling place, which was at my local fire station. The place was crowded—a line stretched right through the station. The line moved slowly, but finally I got to a booth and drew the curtain behind me. There at the top of the list was the key race: "President of the United States: George Bush, Republican . . . Michael Dukakis, Democrat . . . Vote for One."

Vice President Bush was the second-ranking member of an

administration that had placed almost insuperable obstacles in our path. I knew that, if he were elected, I would face at least four more years of financial and political difficulties; four more years in the medical wilderness. My choice was easy that day. I pulled the appropriate lever and left the booth.

As I left the polling place, I called Shar. "How's it going?" I asked her. She barely let me get the words out of my mouth. "I think I've got what you need," she said excitedly. I got in the car and headed back to the clinic, trying not to get too excited myself. Shar was good, she had a practiced eye, but still, she could be wrong.

When I got back to our room in the clinic, Shar was standing by the microscope. I sat down and took a look. She was right. Yes, there lay exactly what we had been waiting and hoping for: the midbrain of a seven-week-old embryo—not even quite old enough to be called a fetus. As such, the tissue was nearly transparent and feathery in character. The midbrain was still attached to the hindbrain, with all landmarks clearly recognizable. After all the damaged tissue we had seen over the previous weeks, this seemed like a small miracle. I said, "This is perfect."

Picking up fine forceps and delicate scissors designed for eye surgery, I steadied my hands to begin the dissection of the critical tissue piece containing the dopamine cells. This step required the utmost care. From the midbrain, which measured only about 5 millimeters (two-tenths of an inch) in length, I had to dissect an even smaller area, barely 2 millimeters wide by 3 millimeters long. This was the region that contained the substantia nigra. Although the substantia nigra is as black as its name suggests in an adult brain, the black pigment has not yet formed in a seven-week embryo. Only more subtle landmarks define the right spot. Any misjudgment at this stage would yield tissue that would be completely worthless for our patient and could even be dangerous if transplanted in error.

I made transverse cuts across the midbrain, about 3 millimeters apart. Then I made two cuts on either side of the midline. A tiny brick-shaped block of tissue floated free. That block contained hundreds of thousands of dopamine neurons, I knew, although I could not see them. I kept my eye on the fragment, lest it float away and be lost in the other tissue debris in the petri dish. Then, putting down the instruments, I sucked up the tissue fragment into a small syringe, and ejected it into a new dish filled with sterile salt solution. As our laboratory experiments had shown, repeated rinses reduced the chance that bacteria might be carried forward to the patient. Finally, I put the tissue in a test tube containing cold salt solution. The chilled tissue would be stable for more than a day.

Before we left the clinic, I had one more task to perform. I went into the procedure room to meet the woman who had just had the abortion. She looked to be in her late twenties. I explained who I was, and I told her exactly what we planned to do with the tissue I had recovered. Would she agree to this? Even before having the abortion, she had signed a release for the fetal tissue to be used for research purposes, but our research protocol required that each woman give specific permission for the tissue to be used for transplantation. She willingly gave it, and signed a consent form. Finally, I asked her permission to take a blood sample, and she agreed to that, too. After I had drawn the blood, I thanked her, and went back to where Shar was working.

As it happened, we were able to obtain another usable tissue sample that day, so by the time we drove back to the Health Sciences Center I was beginning to realize that this was probably going to be the day. Things could still go wrong. The women's blood needed to be checked for viruses—hepatitis B and HIV. And we had to determine the blood type of each embryo. Don Nelson's blood type was type O. We had decided we would use tissue only from blood group O, the universal donor. Only 40 percent of the population are type O, so the chances were slightly against us with each embryo considered singly, but favored us between the two of them. If these tests checked out the way we hoped, there would be no going back. Something that for years had been a theoretical possibility was rapidly becoming a reality. In real life, I knew, things don't always work out the way you hope.

I dropped off the blood sample for virus testing at a lab near the university. When I got back to the Center, I went to the lab of my colleague, neurologist Neil Rosenberg, and gave him a small sample of the fetal tissue from both embryos—different fragments from those containing the dopamine cells. Neil had developed the methodology to determine the blood type of fetal tissue and would now run that test. Other portions went to the bacteriology laboratory for testing. Then I called up Don Nelson at his home west of Denver. "Tonight may be the night," I said. "You'd better come over."

It was about 4 P.M. at that point, and it would be several hours before I got the results of the tests. I decided to call on several people who should know that we were about to proceed. I walked into the office of the chairman of the Department of Surgery, Alden Harken. He had not been a member of our advisory committee, but he was aware that a transplant was being planned. He needed to know that tonight was the night. He listened with interest, and then said "Curt, if this works, people will love you for it. If it fails, they'll kill you."

If that was intended to reassure me, it didn't succeed. In fact, Harken really touched a nerve. I knew there were fellow scientists who would love to see this project crash, and me with it. Medical research is very competitive, sometimes even mean-spirited.

Then I placed a call to the chairman of the human subjects' committee, to let that group know we were going ahead. The committee had approved our plans months before, but I wanted to touch base with them, especially after Harken's comment. It turned out that the chairman was out of the country, so I called the vice chairman, Alan Nies, who was also my direct supervisor as head of my Division of Clinical Pharmacology in the Department of Medicine. He agreed with our plan and wished us luck. Finally, I called Kevin Lafferty, an immunologist and director of the Barbara Davis Children's Diabetes Center on our campus. Earlier in 1988, he had begun transplanting insulin-containing pancreatic cells from human fetuses to treat diabetic patients. He was happy to hear we were moving forward. Everyone said "go," so we did.

Around eight in the evening I finally got word that the test results were good: both embryos turned out to be blood group O, and both women's blood had tested negative for HIV and hepatitis B viruses. Since our plan was to use tissue from only one embryo, I chose the specimen that we had obtained right after I came back from the polling station, because that embryo, at seven weeks of age, should contain cells in optimal condition for transplantation; the other embryo was slightly older.

Since everything looked positive, I called Bob Breeze, who would be in charge from here on. He called his chief neurosurgery resident, John Nichols, and they alerted the operating room and CT-scanner teams. Bob and I had planned exactly how to do the operation over many months. Now our plan would be tested. Bob and I met up on the ground floor of University Hospital, and then went to the hospital lobby to find Don Nelson. He was there all right, with Carolyn and an extended family large enough to take up most of the lobby.

Don sat silent and expressionless in his wheelchair. I knew him to be a smart, active, energetic man, but his disease barely allowed him to talk, let alone keep up with the babble around him. He was making writhing movements with his head and neck, and his hands shook unceasingly—all abnormal movements caused by his Parkinson's disease and the drugs used to treat it. He looked exactly what he was—a man who was losing a twenty-year battle with Parkinson's disease. But, just possibly, this evening would mark a turning point.

CHAPTER 8

The Pioneer

It was 11 P.M. on November 8, 1988, when Bob took Don Nelson to the operating room. Don's wife, Carolyn, accompanied us. It was not the best time of day to start an operation that would last nine hours, but we had little choice. In those early days of our research program, we had not yet developed techniques for keeping the fetal cells alive and healthy for more than a few hours. If we waited till the following morning, the tissue might have deteriorated significantly, so we prepared ourselves for a long night.

For Don it was even harder. The rest of us had eaten that day, but Don, on our orders, had eaten nothing since the previous evening, so he was half-starved before the operation began. In addition, he had stopped his medications at noon, so his symptoms were worse than usual. And while the rest of us had a hundred things to keep us busy, Don's only task would be to sit still and endure whatever was done to him. Imagine an intercontinental flight with the seat-belt sign on throughout, no meals, drinks, or movies, and the flight attendants messing with your brain. But Don proved himself an absolute trouper.

The first order of business was for Bob to shave Don's head. Over the years Bob has acquired, among his many other talents, the skills of a boot-camp barber, and within a few moments he had Don looking like a Marine recruit. He collected the cut hair and put it into a plastic bag. Don still has that bagful of hair, along with many other souvenirs of the operation.

Since the next step was to attach some serious hardware to Don's head, Carolyn elected to leave at that point. After wishing Don well, she retired to a waiting room. It was to be a long ordeal for her, too.

The transplant was to be performed stereotactically, just as we had done with the MPTP-treated monkeys. There was a difference, though. The monkeys were operated under general anesthesia, and therefore we could use a holder that gripped the monkey's head by means of rods placed in its ears, under its eyes, and in its mouth. Such a head-holder causes no harm to the monkey, but it would be quite painful if the animal were conscious. In the case of human patients, on the other hand, neurosurgical procedures are often carried out under local anesthesia and that is what we planned to do with Don. The motivation for this was to avoid the risks associated with prolonged general anesthesia, and to allow us to keep a running check on Don's neurological status. Thus it was necessary to use a head-holder that immobilized Don's head without causing pain.

For this purpose, Bob Breeze used a device called a "halo" that he attached directly to Don's skull. As the name suggests, it is a ring that goes around the head. Unlike a saint's halo, which hovers above the head and lacks any visible means of support, Don wore his horizontally at the level of his nose and ears, and it was clamped to his skull by means of four angled pins that projected upward and inward from the ring. The idea of wearing the halo so low down was to keep it out of the way of the procedures that would be done during the operation.

Bob attached the halo to Don's skull by screwing the four pins through the scalp and a short way into the skull itself. Two of the pins went into Don's forehead and two into the back of his head.

Of course, Bob numbed up the sites where the pins were to go in, so Don didn't feel that part at all. Still, when Bob tightened up the screws to immobilize the halo Don was in real pain for a short while. I'm told that the procedure feels like your head is being tightened in a vise, which is not surprising because that is exactly what is happening. Since the time of Don's operation, Bob has made the procedure a little less like a medieval torture: he gives an intravenous cocktail of short-acting drugs that renders the patient stuporous for those few painful minutes. After that, the halo remains painlessly attached to the patient's head.

The key step in stereotactic surgery is to establish a geometric coordinate system that relates the position of structures within the brain to the halo outside the patient's head. The way Bob did this was as follows. First, he mounted on the halo three sets of three rods, each arranged to resemble a large capital N. The three N's were set vertically on the halo at three positions around the head. Then Don, with all this metal attached to him, was wheeled out of the operating room, along the corridor, into the elevator, and down to the basement of the hospital, where the CT-scanner was located.

A CT (computerized tomography) scan or CAT scan is a three-dimensional X ray. Don lay still on a pad while the X-ray machine revolved around his head. The X-ray data were fed into a computer, and a little while later the computer spat out a series of horizontal "slices" through Don's head, starting at the top of his head and going down to the base of the brain. The CT slices showed many of the internal structures of the brain such as the cerebral hemispheres, the ventricles, and the region that we were particularly interested in—the striatum. (More recently, Bob has switched to an MRI scanner to generate these images, because an MRI scan provides a more detailed image of the brain.)

The CAT-scan images showed not only the structures within Don's brain but also the three N assemblies on the outside of his skull. The horizontal position of any brain structure could therefore be read off in terms of its distance to the three N's. The vertical position of any brain structure could also be determined by reference to the N's. That's because, as you slice downward through a vertical N, the position of the middle, slanting rod gradually shifts from left to right. Thus the position of the slanting rod in a particular horizontal image tells you how far up the N you are, and therefore how far above the halo you are. This may sound a bit complex, but the bottom line is that Bob could calculate the precise three-dimensional position of any structure in Don's brain with respect to the halo on which the N's were mounted.

Once these data were entered into the computer, the N-assemblies were removed from the halo and set aside for the remainder of the operation: all that mattered was that the halo remained fixed rigidly to the skull. This left the space above Don's scalp unobstructed for the next stage of the procedure: the opening of his skull. For this Don was wheeled back to the operating room.

While all this had been going on, I was busy in a nearby lab preparing the tissue for transplantation. Shar Waldrop, who had played such a key role in obtaining the tissue at the clinic earlier that day, assisted me. I took the 2- by 3-millimeter fragment of fetal midbrain and broke it up into a coarse suspension by repeatedly sucking it up and expelling it from a fine glass pipette. Fetal brain tissue is loosely held together in any case, and the mild shearing motions induced by this procedure disrupted the tissue into single nerve cells and clumps of a few tens or hundreds of cells. Then I mixed in a few drops of a "buffer" solution, intended to supply the cells with necessary salts and glucose and to keep the pH (acidity) near neutral. The total volume was now about 400 microliters, or about eight drops of a size that might be dispensed by an eyedropper.

We planned to make a total of ten separate needle passes into Don's brain, so I needed to divide the 400-microliter suspension into ten portions. I set up a row of microcentrifuge tubes—these are tiny plastic tubes with conical bottoms, designed to hold small volumes of fluid—and pipetted 30 microliters of the suspension into each tube. All this had to be done under strictly sterile conditions. Then I capped the tubes, put each one inside a larger tube, and placed these larger tubes in an ice bucket to keep the tissue cool.

Shar took small leftover portions of the cell suspension, added two dyes called acridine orange and ethidium bromide, and examined droplets of these samples under a microscope. This allowed her to estimate how many cells were in the suspension. It turned out that there were about 2,500 single cells in every microliter, plus more cells that could not be counted because they were in clumps. Shar could also tell living cells from dead ones, because the living cells had intact outer membranes that did not allow the ethidium bromide dye to enter the cell, so they remained colorless in the presence of that dye. Using this criterion, Shar found that 85 percent of the cells were still alive.

Having prepared the tissue for transplantation, we went to the operating room, where Bob Breeze was ready to commence the operation. It was about 2:30 A.M. Don was lying on the operating table, which had been configured into something like a lounge chair. His shaven head was fixed to the operating table via the halo and brightly illuminated by the operating lights. In addition to Don, Bob, and myself, there were quite a few other people in the room: Bob's chief resident neurosurgeon, John Nichols; an anesthesiologist, a surgical intern, scrub nurses, a surgery technician whose role was to keep a record of the sequence of injections, and another technician with a videocamera who was taping the entire proceedings. There was a definite air of heightened expectation as Bob swabbed Don's scalp with an antiseptic solution, arranged surgical drapes around the top of his head, and injected a local anesthetic along the line where he planned to make the incision.

Neurosurgery patients, though often conscious throughout their operations, do not usually have much idea of what is going on. They do not feel or see anything. All they can hear is obscure professional jargon among the surgeons, interspersed with an occasional "How are you doing, Don?" or the like. But Don wanted to witness the procedure more directly. He could peer out into the room below the surgical drapes and above the halo that went around the front of his head at nose level. Although he could not see the operative field directly, he could see the glass doors of the instrument cabinets that lined the walls of the room, just a few feet away from him. So he asked a technician to adjust one of these glass doors so it caught a reflection from the mirrored surface of the operating lights. By means of this double reflection, Don was able to get quite a nice view of the top of his own head, so was able to watch the entire operation. If he experienced any squeamishness at seeing his own head cut open, he suppressed it in favor of his determination to learn as much as he could about what was being done to him. When I saw that Don was so involved in the operation, I felt doubly sure that we had chosen the right person to be our pioneer.

Bob Breeze took a scalpel and made a long S-shaped incision in Don's scalp. He retracted the scalp to expose an area of skull about the size of a playing card and scraped the membranous tissue off the surface of the skull so that the bone itself was exposed. Now he had to figure out exactly where to open the skull. For this purpose, he attached a calibrated steel arc onto the halo. This semicircular arc extended over Don's head from ear to ear, rather like the handle of a wicker basket. The needle that would be delving into Don's brain would be attached to the arc, and because the arc was etched with calibrating marks and was tightly fixed to the halo, the needle could be precisely targeted to the ten sites in Don's brain. Bob took the long hollow transplant needle and attached it to the arc so that it was pointing at the calculated position of the striatum. Then he lowered the needle so that it touched the skull and made a mark on the skull at that point. This mark indicated where a hole needed to be drilled through the skull.

After marking this spot, Bob removed the entire arc and took an air-powered drill in both hands. After warning Don of the impeding racket, he began to drill through his skull. The drill did indeed make a lot of noise, and clouds of bone dust flew away from the drill bit and began to accumulate on the surrounding skull: an assistant washed this dust away with saline solution. Don felt no pain, although as with a dental drill there was a certain discomfort from the pressure that Bob had to exert.

The drill bit was circular, so it cut out a coin-shaped flap of bone about one inch in diameter. Bob removed the bone flap, then he cut and retracted the *dura mater,* the tough membrane that encloses and protects the brain. We could now see the brain and its remark-able folded structure, the *gyri* and *sulci*—in the right frontal lobe of Don's cerebral cortex. We would be penetrating Don's awake brain. Once again, we hoped it went well.

The next job was to prepare the tissue for the first injection. Using sterile techniques, John Nichols took the ten microcentrifuge tubes out of the larger tubes I had put them in and arranged them in a rack that he placed on a block of ice. Meanwhile, I arranged the hydraulic system that was to deliver the tissue into Don's brain. This consisted of a syringe whose tip was connected, via several feet of thin plastic tubing, to a very long, thin, hollow needle. The syringe was placed in an infusion pump: this pump simply pushed or pulled on the plunger of the syringe, but it did so much more slowly and steadily than one could manage by hand. I first made sure that the syringe, tubing, and needle were full of sterile saline. It was especially important to get rid of any air bubbles, so that the entire system contained incompressible liquid. This way we knew that if the plunger of the syringe moved inward by a certain dis-tance, an exactly equivalent quantity of the tissue suspension would be delivered into Don's striatum.

Now John uncapped one of the microcentrifuge tubes and placed the tip of the long needle into the droplet of tissue suspension at the bottom of the tube. I set the infusion pump into reverse and allowed the syringe to suck most of the tissue suspension up into the needle. Then I turned the pump off. John took the needle out of the microcentrifuge tube and laid it on the sterile tabletop. It was now full of fetal cells and ready to be inserted into the brain.

Because the needle was about one foot long and only a millimeter in diameter, it was too flexible to drive into the brain by itself; it could not be trusted to stay on track. Also, there was a risk that cells would leak out during the process of insertion. Therefore, we planned to pass the injection needle down the barrel of a larger and more rigid "guide tube"—this was the hollow needle that Bob had already used to mark the place where the hole in the skull was to be drilled. This guide tube—it measured 1.5 millimeters in diameter—was attached to the steel arc, and Bob reattached the arc to the halo, so now the guide tube was pointing directly at the middle of the hole in Don's skull. The idea was first to insert the guide tube to the right position within the brain and then to pass the finer needle down the inside of the guide tube. To reduce injury to the brain tissue while the guide tube was being inserted, Bob put a solid, blunt-tipped stylet into it so that the stylet protruded slightly at the bottom. The blunt tip was designed to push tissue aside rather than to cut it.

For several chapters I've been putting off mentioning an important fact about the striatum, which is that it actually consists of two separate portions, known as the *caudate nucleus* (sometimes abbreviated as the *caudate*) and the *putamen* (see the diagram on page 14). These two parts are close to each other, but are separated by a thick band of nerve fibers. The caudate nucleus is a bit higher in the brain—it actually forms part of the wall of a fluid-filled cavity, the *lateral ventricle*—while the putamen is located deeper in the brain and more off to the side.

In general, the putamen is more severely affected by Parkinson's disease than is the caudate nucleus. Nevertheless, we planned to transplant fetal tissue into both structures in Don's brain, though on the right side only. As I mentioned above, we would need to make ten passes to cover the entire striatum: four passes into the caudate nucleus, and six passes into the larger putamen.

Bob and John had calculated the coordinates of all ten needle passes into brain. The computer printouts for all ten were on paper strips taped to the wall of the operating room. It was John's job to call out the coordinates to Bob. The instrument had to be angled differently for each of the ten passes. Once he had reached the target, Bob withdrew the inner stylet, and John handed him the narrower needle that contained the fetal cells. The back end of the needle was still connected to the plastic tubing and thus to the infusion pump. Bob passed this needle down inside the guide tube until its tip extended 15 millimeters beyond the bottom of the guide tube. Of course, he could not see what was going on inside the brain, but from his calculations he knew that the needle tip must now be at the lower margin of the caudate.

It was 3.50 A.M., and we had reached the point of no return. Once we injected the fetal cells into Don's striatum, they could not be withdrawn. But we didn't stop for any last minute soul-searching. I simply flipped the switch on the infusion pump, the syringe began its imperceptible forward motion, the saline solution began creeping along the plastic tubing, and the cells—though we couldn't see them—were slowly pushed out of the tip of the needle and into Don's brain.

We waited one and a half minutes, during which time 3 micro-liters of the cell suspension containing the dopamine cells were injected. Then I turned off the infusion pump. Bob backed up the entire needle assembly by 3 millimeters and we made another injection, and so on until we had made a total of six injections along the same track, spanning the full vertical thickness of the caudate nucleus. This done, Bob backed the needle assembly right out of the brain, and the first needle pass was complete. I breathed an inward sigh of relief that the first set of injections had been achieved with no untoward consequences and that Don was as alert and involved as ever.

Now we had to prepare for the second planned needle pass, which was also in the caudate nucleus but 4 millimeters behind the first pass. John Nichols read off the coordinates for Bob and then opened the second microcentrifuge tube, placed the tip of the thin needle in the cell suspension, and I again sucked up most of the suspension. Bob made the adjustments in the alignment of the guide tube as instructed by the computer, and the second pass was under way.

It's amazing how quickly a procedure that starts off holding you spellbound can become routine and then tedious, especially at four in the morning. What we were doing required careful attention, certainly, but it wasn't challenging intellectually. In fact, much of the time we spent waiting for the infusion pump to suck up or inject the drops of tissue suspension. We talked among ourselves and with Don, who reminded us that he was hungry and that we were running behind our planned schedule. We could do nothing about his first complaint, but to compensate for the second one I set the pump to a higher speed than we started with. At this higher speed, we found that each needle pass took about twenty minutes, including the time for the computer to calculate the coordinates, the filling of the needle, and the pass itself.

We completed the four passes into the caudate nucleus and then reoriented the guide tube toward the putamen. Because we wanted to go through the same hole in the skull as before, but the putamen is situated more to the side of the brain than the caudate nucleus, the computer told us to angle the guide tube about 30 degrees from the vertical. We also had to drive the guide tube deeper into the brain to reach the putamen. In other respects, the six injections into the putamen proceeded just like those into the caudate nucleus.

Every now and then throughout the operation we would give Don a quick neurological examination. This consisted primarily of asking him to answer a few simple questions and to move his fingers and toes. He never had any difficulty responding appropriately. About 5:30 in the morning, however, when we had about four passes to go, Don began to get restless. Bob recognized this as an almost universal phenomenon; a person can hold out in this strange and uncomfortable situation for several hours, but sooner or later he or she will begin to buckle. It certainly didn't help that Don had been without food for thirty-six hours and without medicine for eighteen.

Don's restlessness increased to the point that he began trying to scratch his head—not a good idea, given that a part of his skull had been removed and a needle was in his brain. The anesthesiologist gave Don an intravenous dose of a mild sedative, Versed. Don had in fact been receiving tiny doses of the drug throughout the procedure, but now Don received a large enough slug of Versed to make him lose interest in any scalp itch he might be experiencing.

Finally, at 7:30 A.M., the last needle pass was completed. Now all that remained was to close the head. First, Bob sewed up the cut in the dura mater so that the cerebral cortex was properly covered again. Then he replaced the bone flap—over time it would grow back into continuity with the rest of Don's skull. Then he released the retractors and sewed up the incision in Don's scalp. Finally, he unscrewed the pins that were digging into Don's cranium and lifted away the halo. Aside from the shaven scalp, the stitches, and a bandage, Don looked none the worse for the procedure.

We weren't quite done with him yet, though. We had to take him down to the basement for a second CAT scan. This is a routine procedure after operations of this kind, the idea being to check for evidence of intracranial bleeding or other problems. Don had in fact been complaining of a severe headache after the bone flap was replaced in his skull. Such headaches are usually caused by air that has entered the cranial space during the operation. The CAT did show some air over the surface of the brain and in one of the brain ventricles, and the presence of the air had caused the brain to shift slightly to one side. This phenomenon is common after brain operations and is not harmful. In fact, Don's headache went away after about three hours. After the CAT scan, we took Don to the intensive care unit where he could eat, sleep, and recover from the surgery while being monitored for any ill effects. Carolyn, who had been waiting anxiously all night, was finally able to see him.

Bob and I shared our relief at things going well. I remarked how this operation was a little like losing your virginity—there's a lot of anxiety in advance, but afterward you recognize it was kind of fun. We were all glad for Don.

The rest of us went our separate ways. For all I know, Bob Breeze may have done another day's worth of neurosurgery—he's that kind of guy. My schedule had me doing a breakfast interview with a postdoctoral fellow candidate from Thailand and then listening to her research seminar. Then I checked on Don. He was doing fine, just receiving a little morphine for headache. At 3:30 that afternoon, I headed home and to sleep.

At some point, I remember seeing the morning newspaper, with a headline announcing George Bush's victory in the presidential election. It wasn't a pleasant note on which to end a memorable night, because it meant that we were likely to face four more years of political problems. But the result had been expected, so I didn't let it get to me.

CHAPTER 9

Follow-Up

I returned to the hospital to pay Don another quick visit later that evening. I was relieved to see that he was alert and in good spirits and looking forward to going home, but we decided to keep him for another couple of nights just in case any complications should set in. Then we turned our minds to the task of letting the world know what we had done. We scheduled a press conference for the following day and sent out an "embargoed" press release—that is, the media were not to publicize it until the time of the press conference itself.

The next day, Thursday November 10, was another busy one for all of us. In advance of the press conference, there were several people I wanted to notify directly about the operation. One person I called was neurologist Stan Burns, who was then at Vanderbilt University in Nashville, Tennessee. Stan had been part of the NIH group that had done so much groundwork for this kind of operation; you may recall that he was the person who earned Bill Langston's hostility when he spirited a couple of the California MPTP patients away to Washington. The NIH group had been prevented from carrying out a fetal-cell transplant themselves because of the moratorium—they were wholly dependent on federal funds. As a courtesy, I let Stan Burns know what we had done. And within ten minutes, a story about our operation was on the Associated Press wire. What happened was that a reporter from an Atlanta newspaper happened to call Burns in connection with some other matter, and Burns told him what he had heard from me. I surmise that the reporter immediately filed an AP story about it. Our phones were ringing pretty much nonstop that morning as media people tried to find out more about the story.

Another person I called was Barry Hoffer, here at the university. As I mentioned earlier, Hoffer was a coauthor of one of the two 1979 studies that demonstrated the effectiveness of dopamine cell transplants in rats. I had not previously told him about our intention to do a human transplant, because I suspected he wouldn't maintain confidentiality or might even attempt to obstruct our plans. When I said that we had done the procedure on our first patient two days earlier, Hoffer was decidedly cool. "Oh, was that done at the Rose Hospital?" he asked, referring to a private hospital nearby. "No," I said, "we did it here at University Hospital." With that, he hung up.

I was left wondering what his response might be. I suspect that Hoffer went to the dean's office and complained about our activities because I soon received another volley of phone calls on top of the ones that were coming from the media. Within fifteen minutes, the chairman of the institutional review board, Norbert Voelkel, was in my office and asking me to appear before a special meeting of the IRB as soon as possible.

But meanwhile I had to take care of the press conference, which was scheduled for noon in the chancellor's conference room on the ground floor of the Medical Center. My wife, Nancy, drove over from Lakewood to attend the press conference, and she was in such a hurry that she got stopped for speeding on the way. "This is your fault," she told me as she showed me the ticket.

There were about twenty-five reporters and cameras for half a dozen TV networks at the press conference. Besides myself, Bob

Breeze and Neil Rosenberg represented the transplant team. Neil was responsible for the tissue typing that ensured we gave Don cells from an appropriate donor, and as a neurologist he had also participated in evaluating Don's condition.

We didn't mention Don's name at the press conference. The last thing we wanted was to have him besieged with reporters while he was in the acute recovery phase. In fact, we had taken some other precautions to preserve his identity from prying eyes. For example, we put a fake name on the door to his room. These efforts were fairly successful at preserving his anonymity, but not entirely. One of Don's neighbors, who saw the press conference, figured that we must be talking about Don and actually called him up in the hospital to wish him well.

Although the medical and scientific issues were what we wanted to stress at the press conference, many of the questions were more political in nature—were we flouting the federal moratorium, what effect on federal policy would the operation have, and so on. I could see that I was fated to remain enmeshed in abortion politics for quite a while. I explained our position: we had done the operation in accordance with all regulations and with the consent of the appropriate committees.

That didn't satisfy everyone, of course, least of all the anti-abortion lobby in Denver. A few days after our press conference a group of anti-abortion activists announced their own press conference, which they held in the street outside the lobby of University Hospital. They denounced what we had done.

Meanwhile, I had to get ready for a trip to Canada. The main convention for brain researchers in North America, the annual meeting of the Society for Neuroscience, was to take place in Toronto, starting that very Saturday. Don was discharged from the hospital on the Friday, with strict instructions to take it easy for a few weeks, and I left for Toronto on the following day.

The way it is with scientific meetings, you have to submit abstracts months ahead of time if you wish to give a talk or present a poster. We submitted several abstracts based on research being done in my laboratory, but none dealt with the operation on Don

Nelson since we had not known when it would take place before the meeting. I expected a busy meeting but nothing like the maelstrom that continued in Toronto. For one thing, reporters from a number of newspapers, including the *New York Times* and *Los Angeles Times,* wanted to do extensive interviews with me on the subject of the fetal transplant.

There was also tremendous interest in the operation among the attendees at the meeting, mingled with a lot of concern about whether we should have done the procedure, given the political climate. I didn't get too much criticism to my face, but I did get some of it indirectly when I read the article written by the *Los Angeles Times* reporter, Thomas Maugh II. Maugh wrote that "most of Freed's colleagues, gathered here for the annual meeting of the Society for Neuroscience, are not pleased with him. . . . It is not even jealousy that he was the first in the United States [to do the operation], although hints of the green monster of envy are visible in the statements of others poised to do the procedure. The problem is when he did it." Maugh mentioned the federal moratorium and the fact that an NIH panel was currently considering the ethical issues involved in the use of fetal tissue in medical research. Maugh's account went on: "Although Freed conducted the transplant with private funds, the operation 'was a real slap in the face to the panel,' said neurobiologist Don Marshall Gash of the University of Rochester School of Medicine. 'He flouted the moratorium and jeopardized the panel's chances of coming up with acceptable voluntary guidelines.' "

For his sharpest critique, Maugh cited Barry Hoffer, who he described as an "intense rival" of mine. Hoffer told him: "I think we have to reassure the public that we will adhere to any guidelines and that we, too, obey the law." Irv Kopin, leader of the NIH group that had been frustrated in its own efforts to carry out a transplant operation, also expressed himself critically to Maugh, as did Roy Bakay of Emory University in Atlanta. As mentioned earlier, Bakay did some of the research on transplanting fetal cells into MPTP-treated monkeys, and he has since done transplants in humans. Recently, however, Bakay has gone off in another, much more futuristic direction: he has implanted interfaces in the brains of paralyzed patients that are intended to allow them to control computers with their thoughts.

Anders Björklund, who had come from Sweden to attend the meeting, did not express any criticism of my work, but he did sound a rather pessimistic note concerning his own two patients, who had been operated nearly a year earlier. "The results have not been impressive," he said.

That was the first I had heard about the outcome of the Swedish operations. Although I think highly of Björklund, I wasn't too disheartened by his words. There were many differences between the way we and his group had conducted our operations. He and I spent an hour comparing notes on our procedures. There were a number of places where I thought our procedure might be better than Björklund's. For example, we used specially buffered solutions to store the fetal tissue, whereas he used unbuffered salt water. We had also used a much narrower needle to deliver the tissue than he had, and we had placed the needle tracks closer together. Minor though many of these details might have seemed, it was possible that they were the key to success or failure.

Of course, Maugh also reported my response to the criticisms. "When the government restricts scientific research," I said, "even though it doesn't prohibit it, that casts a pall over the entire research community. People may be upset with me now, but in six months, everyone will want to know what happened to my patient." This is in fact exactly what happened.

It is true that an NIH panel, headed by retired U.S. District Judge Arlin Adams, had been considering the ethical issues over the previous few months. At the time we operated on Don Nelson, however, I already knew what the panel's recommendations were to be, even though they were not officially released until early the following year. They proposed that the medical use of fetal tissue be permitted if a number of conditions were fulfilled, most notably that a woman's decision to have an abortion be kept completely separate from any decision about donating the resulting tissue. We adhered scrupulously to that condition with Nelson's transplant and all subsequent transplants; the women signed documents consenting to the abortion before any mention was made of a possible tissue donation. In Nelson's case, in fact, the woman wasn't approached till the abortion had already been completed.

It so happened that one of the events at the Toronto meeting was a panel discussion devoted to research ethics in neuroscience. The discussion was chaired by Fred "Rusty" Gage, a neuroscientist who has himself been involved in transplantation studies in animals. (He is now at the Salk Institute in San Diego.) Although fetal-tissue transplantation wasn't a scheduled topic of discussion, it did come up, in a big way. Panelists and audience members expressed quite a diversity of opinions on the matter, but few if any speakers thought that the medical use of tissue derived from abortions was unethical.

I gave a brief account of the transplant we had done on Don Nelson, carefully explaining how we had followed all regulatory procedures. That was too much for Barry Hoffer. He leaped up and demanded that "we should obey the law." Hoffer's comment irritated me because, in fact, I paid scrupulous attention to all regulations and had received permission from the chancellor, the hospital director, and the human subjects committee.

Meanwhile, a glitch had occurred back in Denver. On Sunday morning, I got a call from my lab. Some of the leftover cells from the tissue we had implanted in Don Nelson's striatum had been tested for the presence of microorganisms, and one of the samples had grown out a potentially harmful bacterium. I discussed this by phone with Bob Breeze. We agreed that the sample had most likely become contaminated after the remainder of the tissue had been transplanted into Don's brain. But given that a brain infection can have such devastating consequences, we decided we should act aggressively to deal with the small possibility that we had caused one. So I called Don, explained what had happened, and asked him to come back into the hospital for a few days. He came back in that same day, and Bob performed a spinal tap to get a sample of cerebrospinal fluid. This sample did not show any evidence of infection. Still, Don was placed on intravenous antibiotics for three days, and he went home the following Tuesday, a week after his transplant operation.

A day or two after I got back to Denver, I called Don to see how he was doing. Carolyn answered the phone. "He's outside shoveling snow," she told me. I was quite disconcerted to hear this. We had told him to avoid any kind of strain for a few weeks after surgery, and here he was doing something that would be a strain for a healthy individual, let alone for someone with Parkinson's disease who had just had brain surgery. I asked Carolyn to tell him to stop. Still, I was pleased that Don had recovered sufficiently from the rigors of the surgery that he would want to resume physical activities.

Over the ensuing weeks and months, our main efforts were devoted to monitoring Don's condition. Initially, Neil Rosenberg went to Don's home about three times a week at 6:30 in the morning, so that he could evaluate Don before and after his morning medications. Also, Don continued the computerized testing with the equipment we had set up in his home. In addition, we had him come in as an outpatient from time to time to test him more extensively, adjust his medications, and so on. I doubt that a person with Parkinson's disease has ever been subject to such thorough and time-consuming monitoring. Don took it all in good humor.

We did see a gradual improvement in Don's condition over the months after the transplant. In terms of the computerized testing, we saw a significant increase in his speed of finger movements within two months after transplant. As was to be expected, the improvement was only in his left hand—the hand controlled by the right side of the brain, where we had placed the fetal cells. Don's walking speed also improved, although this was true only after his morning drugs; tested before his morning dose, Don actually walked slower during the first year after surgery than he had before.

Don also noticed a gradual improvement in his abilities to carry out the activities of daily living. His speech improved, and he regained his ability to whistle—a favorite pastime of his. He fell less often and walked more confidently, relying less on his cane or crutches. He had fewer episodes of "freezing." Eating became a lot easier as he regained better control over his knife and fork. He could tie his tie for the first time in years. His chronic constipation resolved.

Because of his improved state, Don embarked on several projects at home. In April of 1989, five months after his surgery, he and Carolyn repainted and wallpapered a bedroom, and he replaced the paneling on the walls of their family room. In the summer of 1989 he became more active in the yard, digging up bushes and trees and planting flowers. The lawn mower still defeated him, because it insisted on moving forward faster than he could follow, but in general he could keep the yard in order. In September of 1989, Don and Carolyn traveled to Michigan to attend a wedding, and he found himself able to mingle with 250 people at the reception without a problem.

Given these improvements, we wondered whether we could reduce Don's medications. At the time of the transplant, Don was taking 900 milligrams of L-dopa (in the form of the L-dopa/carbidopa combination, Sinemet), along with three other drugs. Over the course of the first year after the transplant, we experimented with reducing the various medications, but without a great deal of success.

Part of the reason for this had to do with the fact that we made a transplant only into the right side of Don's brain. By doing so, we had introduced a functional asymmetry: the left side of Don's body was now controlled by a striatum that had had its dopamine supply boosted by the transplant, while the right side of his body was still controlled by a striatum that was severely deficient in dopamine. Thus, if we tried to reduce L-dopa, the left side of his body did well but his right side became stiff and slow. If we increased his L-dopa, on the other hand, his right side did well but his left side showed signs of dopamine overdose, such as dyskinesias. We realized that it would be easier to adjust the medications of someone who had bilateral transplants.

All in all, Don was significantly improved a year after surgery, with much of that improvement occurring within the first three months. He was far from cured, though. He still walked slowly compared with a healthy individual. He still had problems turning in bed and in swallowing his food. His handwriting was still illegible. He remained quite obviously a person with Parkinson's disease, even if some of the disease's worst burdens had been eased.

Nine months after the transplant, we wanted to carry out another PET scan, to see if we could find any evidence of the survival of the grafted dopamine cells in Don's striatum. Unfortunately the New York facility where Don had been PET-scanned prior to his surgery had closed down, so we had to send him to another facility, at the University of California, Los Angeles. No two PET machines give identical data, however, so we couldn't really tell with any confidence whether there was improved uptake of fluorodopa—the tracer used to monitor the functional status of the dopamine system—in Don's right striatum, compared with the previous scan. Still, the scan did provide a benchmark with which we could compare further scans a year or two later.

Reasonably encouraged by the results of Don's transplant, we started to plan for a second operation. We had been approached by "Anthony Marsh,"* a trial attorney. He was in his early sixties and had had Parkinson's disease for about twenty years. The disease destroyed his marriage, and it also impoverished him, as he had to give up his law practice. Still, he was an accomplished man who was trying to start a new career as a writer.

There were two things a bit unusual about Marsh's condition. One was that he didn't respond to L-dopa well. Don Nelson, like many other people with Parkinson's disease, would freeze up if he stopped taking L-dopa for more than a few hours, but we could take Marsh off L-dopa for an entire weekend before we saw much of a worsening in his condition. In retrospect, it may be that this unresponsiveness should have told us that he would not be a good candidate for a transplant. Another unusual aspect of his case was that he had already been subject to brain surgery for his illness:

*Names indicated by quotation marks at first mention are pseudonyms.

neurosurgeons had destroyed part of a brain region called the thalamus on both sides, in order to alleviate his tremor. This so-called thalamotomy operation had had some success in relieving Marsh's tremor, but it introduced an unknown factor into the question of how he might be affected by a transplant.

Counterbalancing these considerations was Marsh's strong desire to have the transplant surgery. We therefore decided to go ahead, and performed the transplant in January of 1990, about fourteen months after the operation on Nelson. Although Marsh was about equally affected on both sides of his body, we decided to transplant fetal cells only into one side of the brain. We were simply not yet comfortable enough with our procedures to venture a bilateral transplant. We made the injections into the caudate nucleus and putamen on the left side.

We did make a couple of technical changes that greatly speeded the surgery. For one thing, Bob Breeze had designed a device that eliminated most of the calculations required to aim the injection needle. With Don Nelson, the coordinates for each penetration had to be calculated separately. After Nelson's surgery, however, Bob made a metal template drilled with a row of holes at 4-millimeter intervals that acted as a guide for the injection needle. At the beginning of the surgery Bob placed the needle in one of the holes and aimed it at the center of the putamen, and he rotated the plate until the row of holes was aligned with the long axis of the striatum. Then, after making the first injection, he simply had to move the needle to the next hole in order to get the 4-millimeter spacing of the injections that we needed.

Another change we made was to inject the cells by hand pressure on the syringe, rather than letting an infusion pump do the job. It turns out that an injection in the form of a quick "bolus" is more effective in getting the cells out of the needle than a slow infusion is. The reason is that with a slow infusion the fluid streams out of the needle but the clumps of cells can get hung up inside it. Between these two technical improvements we were able to cut the duration of Marsh's surgery by about half, compared with Nelson's.

Another important difference between the two transplants was that, unlike with Nelson, we treated Marsh with immunosuppressants. Since we did not know whether such treatment was necessary or not, we decided to immunosuppress about half of our first series of patients and see if there was any difference in the clinical outcome. We treated Marsh with high doses of steroids as well as with cyclosporine.

The operation on Anthony Marsh went off uneventfully, but a day or two later he was found wandering around the hospital in his hospital gown, not knowing exactly where he was or why. We got him back to his room in short order, of course, but it was about a week before his mental status returned to normal. At the time, we blamed his confusion on the steroids he was receiving—they certainly can have peculiar mental effects in the high doses that are necessary for immunosuppression. Subsequently, we observed short periods of confusion, lethargy, or memory impairment in other patients after transplant surgery, even if they had not been given immunosuppressant drugs. We came to realize that making a large number of needle passes into the brain can compromise its function during the week or so after the operation.

Although Anthony Marsh's confusion cleared up, we did not see any improvement in his Parkinson's disease during the months after surgery. The transplant seemed to have no effect, either for better or for worse. This was obviously a disappointment to us, but given the apparent benefit in Don Nelson's case, and the lack of any harmful effects with Marsh, we felt well justified in continuing the study. About a year after Marsh's surgery we prepared for a third transplant. Meanwhile, Marsh moved to Costa Rica, where, in spite of his continuing illness, he managed to write a couple of novels.

Our third patient was another Denver-area resident, Robert Majzler. I first met him at a talk I gave for a local Parkinson's disease support group; after my presentation he told me how keen he was to volunteer for the transplant surgery. Majzler was an ex-Jesuit priest who had married an ex-nun, and they had settled down in the Boulder area to start a family. Unfortunately, Bob was diagnosed with Parkinson's disease around the time that their first child was born, in 1981. Initially, his only problem was with his right arm, which became sluggish and clumsy and refused to swing when he was walking. He didn't have much of the typical Parkinsonian tremor, however.

Over the following few years the disease advanced rather rapidly, interfering with Majzler's capacity to work or even to be an active parent. His left side became involved, and in spite of being on L-dopa he began experiencing frequent and prolonged "off" periods when he was largely immobile. His oldest son recalls how, when he was only six or seven years old, he had to pull his father out of bed. Within a few more years, Majzler was spending his days in a chair in his living room, a prisoner in his own body and entirely dependent on his family for personal care.

After I met Majzler at the Parkinson's disease support group, I had him come into the clinic for a neurological examination. The rapid fluctuations in his condition were impressive. When he entered the room he was wildly dyskinetic—limbs and head flying in all directions. Then, toward the end of the examination, he froze up completely. I had to dress him myself to get him out of the door. I accompanied him out of the hospital, and he managed to walk that fifty yards' distance—backwards! That's another peculiar thing about Parkinson's disease: patients who can't take a step forward can sometimes walk backwards quite well. It's such a common phenomenon that it's been given a medical name: *retropulsion*. It does attract stares, but Parkinson's disease patients have to deal with that much of the time anyway.

Between Marsh's and Majzler's operations we made significant changes in the way we handled the fetal tissue. Most important, we developed techniques for keeping the fetal tissue alive in tissue culture for several days. This meant that we could accumulate tissue from many fetuses to transplant into one patient, and it also meant that we could schedule the operations ahead of time and perform them during normal daytime hours.

Another technical change involved the preparation of the tissue for implantation. Instead of breaking it up into a coarse suspension of cells and cell clumps, I placed the original piece of brain in a pipette and extruded it from the tip rather in the way that spaghetti is extruded from a pasta machine. The resulting "noodles" of tissue could then be sucked up into the injection needle and deposited in the brain during the withdrawal of the needle. This laid the "noodle" down as a uniform strand within the striatum.

Yet another change we made involved the locations of the injections. We finally resolved to make bilateral injections, but at the same time we decided not to inject the front division of the striatum, the caudate nucleus. As I mentioned earlier, the caudate tends to be less affected by Parkinson's disease than the putamen. In Bob Majzler's case, in fact, the PET scan showed that the caudate still had dopamine, whereas the putamen was grossly deficient. Thus with Bob, and with all subsequent transplants, we made injections into the putamen only, but on both the left and right sides.

In spite of the fact that we were injecting both sides of the brain, our procedures had become so efficient that the operation lasted only about three hours, compared with more than ten hours for Don Nelson. This, and the daytime scheduling, greatly relieved the stress on all participants, most especially on the patient himself.

Although he suffered no ill effects from the surgery, Bob Majzler's condition worsened during the weeks after the transplant. It was evident that making so many needle passes into the diseased striatum had compromised whatever function it still retained, making it even more difficult to control his symptoms with drugs. Majzler's worsened condition put a tremendous strain on his wife, Helen, who had to do everything for him, from helping him eat to moving him around in bed during the night. Helen is a strong woman, but she is only five foot two inches tall, whereas Bob is over six feet. I realized the toll it was taking on her when she brought Bob in for an outpatient visit a few weeks after the operation. While I was examining him, Helen lay down on an examination table and fell asleep. He was alert; his spouse was exhausted.

To our relief, the worsening of Majzler's disease status lasted only a few weeks. And, even more pleasing, we began to see an actual improvement in his condition compared with his preoperative state. Already, at six weeks after surgery, the computerized testing of his walking speed and his hand movements showed an improvement. After a few months, both his dyskinesias and his "off" episodes decreased in number and severity. Sometimes he appeared nearly normal.

Bob Majzler took advantage of his improved condition to do many things he hadn't been able to do for years. Like Don Nelson, he undertook major redecoration in his home. His oldest son had just graduated from high school, and to mark that accomplishment Bob redid his room, working six-hour days for three weeks. Then he did the same for his other son. Bob also began going to the movies again, as well as to his sons' sports activities. He even started riding a bicycle, but switched to a tricycle for the sake of the extra stability. Again like Don Nelson, he and his wife took a trip back east to visit relatives, and, as he put it, he "didn't miss a step" during the trip. Perhaps most important for Bob, though, was that he and Helen were able to go to church together again. About a year after the operation they attended an Easter service, Bob's first for quite a few years. Helen commented, "I guess you could call it a resurrection."

We were much more successful in reducing Majzler's drugs than we had been with Nelson. Already by a year after the operation we had reduced his L-dopa doses by 39 percent, and we made further reductions later, though we were never able to take him off the drug completely. His condition continued to oscillate with his drug schedule: shortly after he took his L-dopa he could walk quickly and smoothly, but as the time for the next dose approached he would become slow and stiff. The radical on-off fluctuations, though, were a thing of the past.

The first three transplant surgeries were paid for out of the money donated by the Stanton brothers, but after that the money ran out. In November 1989, the Bush administration rejected the recommendations of the NIH's committee and made the fetal-tissue moratorium permanent. Now federal support was out of the question, and we could raise only modest amounts of money from private sources. We were in a quandary as to how to finance future transplants, since the operations and the associated medical costs were running at about $30,000 per patient.

This problem is actually a rather common one, even when the touchy issue of fetal tissue is not involved. The NIH typically want strong evidence that a procedure is effective before it will fund research on that topic, so the initial experiments to obtain that evidence usually have to be funded from some other source. With research into new drugs, the pharmaceutical companies are a major source of funds. With a surgical procedure like ours, though, no commercially exploitable product is being developed, so neither industrial support nor venture capital is forthcoming.

To my mind, the medical insurance industry should be more involved in funding medical research. The potential savings to the industry from a new and effective procedure are enormous, particularly with progressive, debilitating conditions like Parkinson's disease. Just curing one patient at an early stage of his or her illness could save an insurance company several hundreds of thousands of dollars, and there are a million or so people with Parkinson's disease in the United States. It's optimistic to think that any new and untested treatment is going to be an effective cure—many will turn out to be ineffective or no better than current therapies. Still, the insurance industry maintains too conservative a stance, in my estimation. Not only do they invest very little into medical research, they also pay out very little to patients for treatments they deem "experimental."

Anyway, it became clear that there was only one way for us to finance the continuation of our program, and that was to bill the patients themselves for their transplants. None of us felt especially comfortable with that decision, since it meant that our patients were paying to be research subjects, but it was just a reality: either charge patients or abandon the program.

The first patient who volunteered for a transplant on this basis, the fourth in our overall series, was a sixty-four-year-old woman from Alabama named Faye Day. She was also the first woman to apply who was a good candidate for a transplant. In general, men are more willing to try new and risky activities than women are, and Day was the only woman to buck that stereotype until we were four years into the program, when we started getting inquiries from about equal numbers of men and women.

Day's most serious problem was her walking. She moved in a painfully slow and shuffling fashion, and she fell constantly—often as many as a hundred times in the course of a day.

We were able to help Day get some financial support from her insurance company—they paid about half of her medical expenses, which was enormously helpful to her since she wasn't a rich woman. We used the same procedure as with Majzler, and she recovered fairly well from it without too much worsening of her condition. Then, starting about six weeks after the transplant, her walking began to improve and she fell less. Soon she was falling no more than once a month.

Day was PET-scanned before and after the operation by Dr. John Mazziotta at the UCLA facility. There was a dramatic improvement in the PET image after surgery: the putamen, which took up very little fluorodopa before the transplant, lit up as strongly as the caudate nucleus afterward. She was our first patient who showed unambiguous evidence of survival of the grafted dopamine cells in her PET scan.

Day was operated in July of 1991, six months after Majzler. Now, however, we began accelerating the program. This was partly because we felt we had established techniques that worked, so we no longer needed to reevaluate everything after each patient came through. In addition, applicants for fetal-cell transplants were becoming more plentiful, in spite of the financial burden that the surgery imposed on them. This was mainly because of the national media attention that our transplant program had attracted. In early 1992, for example, *60 Minutes* ran a report about our program; it showed Faye Day walking confidently, less than a year after her operation. We received over a thousand telephone inquiries about our program after the show aired.

Just a month after we carried out the transplant on Faye Day, we carried out our fifth transplant, on a fifty-eight-year-old man from California named Robert Orth. He was probably in worse shape than any patient we had operated on before. He had the most extreme on-off fluctuations: at one point he would be racked with violent dyskinesias, such that he might throw himself bodily from his chair onto the floor; at another point he would be totally locked in, unable to move or speak or participate in the world in any way. There was simply no middle ground. When he could speak at all his words were incomprehensible, except perhaps to his wife, Chris, and he avoided any kind of public outings if he could help it—the embarrassment was too great. About three years before his transplant, when he was not quite so disabled, he had gone to the Department of Motor Vehicles to renew his driver's license: the clerk took one look at him and confiscated his license on the spot. This indignity, more than anything else, was the spur for him to sign up for a transplant operation.

Orth had worked for the phone company, and when he became too disabled to work the company was helpful in ensuring he got a good early-retirement plan. Still, the Orths were far from affluent. Although neither Dr. Breeze nor I charged for our services, the Orths ended up having to pay about $45,000, including their travel expenses, and none of this was reimbursed by insurance. They went seriously into debt and had to sell the double-wide trailer in which they lived. They moved into a much more modest trailer in another park, perched among the sand dunes on the Central California coast. Luckily, they really liked their new location and have stayed there ever since. Still, it was a vivid illustration to me of exactly what a $45,000 expense can mean to a couple already coping with the financial burden of a serious disability.

Bob Orth did well after the transplant. His dyskinesias and his "off" episodes abated, and his speech improved dramatically. He even regained his driver's license. I've come to realize that the patients who improve after transplant surgery go through a kind of second adolescence, one of the main symptoms of which is an irrepressible urge to get behind the wheel of a motor vehicle. Like Nelson and Majzler, he set off for points east. His wife, who was totally exhausted by all the turmoil of Bob's illness and the transplant, refused to go with him. "You got a magic pill and a magic operation," Chris told him. "Do you think that I got a magic pill, too? I didn't."

We performed transplants on three more patients in 1991, five in 1992 and another five in 1993. Some of the patients were markedly improved by their transplants, some showed mild improvement, and some were essentially unchanged. It was difficult to tell what factors decided whether a particular patient would be helped or not. There was a clear suggestion that age played a role, though; the younger patients—those under sixty—seemed to do better after surgery than the older ones. Even here there were exceptions: Faye Day, for example, was sixty-four when she had her surgery but she was greatly helped by it.

Each one of the patients in those early years is indelibly etched in my memory: "Vince Ahlberg," who became so anxious during the surgery that we had to stop in the middle and complete the procedure two weeks later; Anderson Whitt, who bought a new car after his operation and totaled it, but walked away unscathed—I can rattle off their histories and am familiar with all their strengths and foibles. I saw each one of them repeatedly before and after their transplants, and I still see many of them today. We form an extended family, united by our memories of the pioneering days of fetal-cell transplant surgery.

During these years other groups were also performing fetal-cell transplants for Parkinson's disease. Eugene Redmond's group at Yale, who had also managed to obtain some private funding, did their first transplant only about a month after ours, and they did several more during the following three years. Several aspects of their technique did not bode well for success, however. For one thing, most of the transplanted tissue was derived from fetuses that were likely too old—ten or eleven weeks of fetal age—to contain dopamine cells at the optimal stage for transplantation. In addition, they had frozen the fetal tissue in liquid nitrogen. The freeze-thaw cycle is a brutal experience for nerve cells; when we tried this technique, up to 90 percent of the cells were killed. Thus I was not entirely surprised to learn that Redmond's patients experienced little benefit from the transplants, and in one patient who later died no surviving dopamine cells were found in the striatum at autopsy.

More promising news came out of Sweden. Although Björklund's group had little success with their first two transplants, they did fare better with later patients. Of particular interest were two patients operated in 1989: these were none other than George Carillo and Juanita Lopez, two of the MPTP-poisoned drug users who had been under the care of Bill Langston in California (see chapter 5). In 1982, Carillo and Lopez had responded well to L-dopa; the first dose of the drug had transformed them from mute, frozen statues to nearly normal human beings. But this happy state of affairs did not last long. Within a year, both were experiencing severe side effects of the L-dopa treatment: Carillo was hallucinating, and Juanita was racked by dyskinesias so violent that, like Robert Orth, she often threw herself from her chair to the floor. It probably didn't help their condition that both of them spent much of their time in prison for drug-related offenses; on one occasion, Carillo was brought to Langston's hospital with multiple stab wounds he had suffered in a prison brawl.

Many a physician would have given up on such troublesome patients, but not Langston. These individuals had a unique kind of Parkinson's, and Langston had an abiding desire to help them. When he heard about the fetal-cell transplants, it occurred to him that the MPTP patients might respond better to a transplant than the usual victims of Parkinson's disease. His rationale was as follows: the cause of their Parkinson's disease—the MPTP itself—had long since been washed out of their systems. Therefore, new dopamine cells implanted in their striatum would not be exposed to the same factor that had caused the death of the first set. With most Parkinson's disease patients, on the other hand, there was the possibility that whatever process had killed the patient's own dopamine cells would attack the transplanted cells too. This issue was discussed at the New York meeting arranged by Björklund in 1986 (see chapter 6). Even today, it hasn't had an unequivocal answer, though we do know that transplanted cells can survive for many years in regular Parkinson's disease patients.

At any event, Langston moved heaven and earth to arrange for Carillo and Lopez to receive fetal-cell transplants in Sweden. And they did both improve as a consequence of the transplants they received. By two years after surgery their symptoms had greatly improved and their L-dopa doses had been reduced by about two-thirds. Juanita Lopez in particular looked much like a normal person.

In 1992 the Swedes, the Yale group, and ourselves published three back-to-back papers in the *New England Journal of Medicine* that summarized our findings to date. This event was a bit of a milestone for the fetal-tissue transplant procedure. Although the results varied among the three groups, the findings as a whole were reasonably positive. Among our own patients, about two-thirds had experienced some degree of improvement. The publication of these papers legitimized the procedure in many people's eyes and made it easier to persuade insurance companies to contribute to our patients' expenses.

Yet the studies raised more questions than they answered. Which patients were most likely to be helped by the procedure? What was the best way to handle and inject the fetal tissue? Was immunosuppression helpful or harmful? Because the various studies had been done in such different ways without any systematic comparisons of techniques, we could not be sure of the answers to any of these questions. Even the most basic question—did the procedure benefit patients?—had not been answered with complete certainty. There was still at least a theoretical possibility that the benefit that we and others had observed was a purely psychological effect (a "placebo effect") or was a nonspecific effect of disturbing the organization of our patients' brains. To begin to answer these questions would take a much larger and more elaborate study than any of our groups could mount without government help. As another presidential election loomed, we renewed our hopes that such help might be forthcoming.

CHAPTER 10

Politics, Continued

The efforts to change federal policy on fetal-tissue research began long before the 1992 election. Both the Senate and the House of Representatives had Democratic majorities, after all, so there was considerable sentiment in Congress in favor of an attempt to overthrow the Republican administration's moratorium, which had been made permanent in November of 1989, just about a year after Don's operation and Bush's election to the presidency.

In August of 1990, Representative Henry Waxman of California introduced the Research Freedom Act. The purpose of the bill was to overthrow the administration's moratorium on the funding of fetal-tissue transplantation research by directly authorizing the NIH to support such research. It also established a variety of procedures to prevent abuses of the research, such as the requirement that a woman give consent for an abortion before she was asked for consent for medical use of the fetal tissue.

In introducing the bill, Waxman emphasized the potential medical benefits to be gained from this line of research. He also poured scorn on the idea, espoused by the administration, that permitting the funding of research into the transplantation of fetal tissue would somehow increase the number of abortions. "The [administration's] policy is comparable to one of banning lifesaving organ transplants to limit traffic accidents, and it is equally ludicrous."

The bill didn't make much progress in 1990, but in the following year it was reintroduced, and in July 1992 it passed the House of Representatives by a vote of 274 to 144. This was very nearly a two-thirds majority, which raised hopes that an eventual veto by President Bush could be overridden.

The Senate version was introduced by Senators Edward Kennedy of Massachusetts and Brock Adams of Washington. Kennedy was chairman of the Senate Labor and Human Resources Committee, and Adams was a member of the same committee and chairman of its Subcommittee on Aging. Thus, the bill was rapidly shunted to the Labor Committee for hearings.

In early November 1991 I got a call from a staff member on Kennedy's committee, who asked me to testify at the hearings. I did so later that month, as did several men and women with Parkinson's disease and other diseases likely to be helped by research using fetal tissue. I told the committee about the progress we had made, our hopes for developing an effective treatment, and the problems that had been caused by the funding moratorium.

The senators may have been particularly disposed to listen sympathetically because one of the most respected members of Congress, Representative Morris K. Udall of Arizona, had had to resign six months earlier on account of Parkinson's disease. Mo Udall had served in Congress for thirty years and was particularly known for his efforts on behalf of the environment. As his disease progressed, he became liable to falls, and one of these resulted in severe brain injuries from which he never recovered. He died in 1998.

At the hearing, Mo Udall's daughter Anne talked of her father's fifteen-year struggle with the disease, and she revealed that no less than ten members of her family had the same condition. Although most cases of Parkinson's disease are sporadic, there is a form of the disease that runs in families, and the study of these families has recently led to important discoveries concerning the causes of the disease, which I'll discuss in chapter 20. At the Senate hearing, however, the main point of Anne Udall's testimony was to bring home the personal impact of the disease, and she accomplished this in a very moving way.

Another person who testified at the hearing was Joan Samuelson, an attorney from California who had been diagnosed with Parkinson's disease in 1987. In 1991 she founded the Parkinson's Action Network, an advocacy group, and devoted herself to lobbying for the overthrow of the fetal-tissue ban and for increased funding for Parkinson's disease research. Joan combines a powerful personal motivation to conquer the disease with great political savvy and organizing skills. She was really the organizing force behind the Senate hearings; she had personally lobbied a large number of senators on behalf of the bill. In her own testimony she described what it was like to gradually lose control of her body, to know that research should be able to find a solution, and yet to see politicians trying to obstruct instead of solve the problem.

Kennedy's committee voted in favor of the bill and sent it back to the Senate floor, where it came up for debate in early 1992. During this time I was taking a sabbatical from the University of Colorado and was in Paris learning molecular biological techniques. While in Paris, I received word that some Republican senators were considering voting against the president on the issue of the fetal-tissue ban. In particular, Senator Connie Mack of Florida, a longtime opponent of abortion rights and a stout ally of President Bush, was undecided about which way to vote and was interested in learning more about the potential benefits of fetal-tissue transplantation. After discussing the issues with his assistant over the telephone, I wrote Mack a letter that laid out the case for the Freedom of Research Bill in as persuasive terms as I could muster:

> I understand how difficult it must be to balance the needs of your Florida constituents, many of whom are elderly and who suffer from diseases that fetal tissue

research is likely to help, and your opposition to abortion. Abortion is not the issue. Nothing in this bill will change anything about abortion in the United States. It is a bill to support research using fetal tissue that is otherwise thrown away.

I described our research program and mentioned how several patients had benefited markedly from the fetal transplants. "A California man had lost the ability to communicate because his speech was so distorted by the effects of the disease," I wrote, referring to Robert Orth. "His driver's license had also been revoked because of his poor control of movement. Since surgery in August, his speech has returned, he has passed his driver's test and is now driving around the United States celebrating the recovery of his body."

I mentioned the importance of fetal tissue in earlier medical breakthroughs, such as the development of a polio vaccine, and pointed out the irony that federal funds were supporting animal research using human fetal tissue: "Federal grant funds can be used to put aborted human fetal tissue into a rat but not into a human. This is irrational."

I also pointed out that the Freedom of Research Bill included safeguards to prevent abuses in the collection of fetal tissue, and that if the bill was not enacted, there would be no legal safeguards to prevent such abuses in privately funded programs, concluding:

Parkinson's disease costs billions of dollars per year in medical care in addition to the cost of lost income from working lives cut short. The research to define the value of brain repair with fetal tissue will cost a few million dollars. On all grounds—humanitarian, economic, and scientific—there is a need to know whether repair of the human brain is possible. As you know, your vote can be decisive for passage of this bill. I urge you to provide your critical support and vote for the Research Freedom Act.

In April of 1992 the full Senate voted in favor of the bill by an 87–10 margin. Particularly striking was the fact that several senators who were outspoken opponents of abortion voted in favor of the bill: these included Senators Mark Hatfield, Bob Dole, John Danforth, Strom Thurmond—and Connie Mack.

Shortly after the vote, we were contacted by President Bush's son Jeb, who was an up-and-coming politician in Florida years before becoming governor. He asked for a copy of the letter I had written to Senator Mack, saying that the president wanted to read it. I was of course flattered that the president wanted to see what it took to persuade one of his senators to vote against him.

Whatever President Bush may have thought of my letter, he did veto the bill. With the votes as they were, however, Bush realized that an override of the veto was a distinct possibility—something that would have been an embarrassing political defeat for him. He therefore attempted to undercut some of the support for the bill in the House of Representatives by earmarking $2 million to establish a "fetal-tissue bank." The aim of this bank was to collect fetal tissue from spontaneous abortions and from abortions that were medically necessary to save the life of the mother (as for example in the case of ectopic pregnancy, where the embryo implants in the fallopian tubes).

The administration released some letters from doctors in support of the idea of a fetal tissue bank. Robert Cefalo, an obstetrician-gynecologist at the University of North Carolina who had been a member of the 1988 panel that reviewed the ethics of fetal-tissue transplantation, wrote: "There is evidence that a proportion (5% to 7%) of spontaneous miscarriages will provide tissue appropriate for use in tissue transplantation research. There is a need to confirm this data and . . . a collaborative network of transplantation researchers . . . could accomplish this."

Although I and other experts had testified that the proposed fetal-tissue bank was not a feasible strategy for obtaining usable fetal tissue, Bush's action did convince a handful of wavering representatives that the Freedom of Research Bill might not be necessary, or that action on it should be postponed. Joan Samuelson labored hard and long to get a two-thirds majority for the veto override, but her one-person lobbying drive was unsuccessful: the override failed in the House by 14 votes.

The fetal-tissue bank was in fact set up, and over the next year or two the remains of 1,200 fetuses, most of them derived from spontaneous abortions, were collected and examined. The results of this study appeared in January 1995, long after George Bush had been voted out of office. Just seven of the 1,200 fetal samples yielded usable tissue. Of the remaining samples, most consisted of dead tissue, because spontaneous abortion usually results from the death of the fetus within the uterus. Other samples contained tissue that was chromosomally abnormal or that was contaminated by bacteria. At a total cost of $2 million, the seven usable samples cost nearly $300,000 apiece. It was galling to think that so much of the public's money—enough to fund about sixty fetal-tissue transplants—had been thrown away in the name of saving President Bush from political embarrassment.

As an interesting postscript on this episode, I was at a dinner in Washington in April 1998 and met with Louis Sullivan, who had been Secretary of Health and Human Services during George Bush's presidency. I mentioned our disappointment with Bush's veto of the Research Freedom Act. He remembered this issue clearly and said, "You know, the president wanted to support you, but the politics wouldn't allow it. He had no choice but to veto that bill."

CHAPTER 11

A Warning

In 1992 I learned of a bizarre autopsy that had taken place the previous year in Boston. The autopsy was of a Parkinson's disease patient who had received a fetal-cell transplant in China, and the findings were so disturbing that they briefly threatened to impede the continuation of this kind of research. Over the following years, I have found out more details of this story, which speaks volumes about what can happen when human desperation encounters medical misadventure.

The story mainly concerns two people, a well-known distance runner named Max Truex and a neurosurgeon named Robert Iacono. Truex was born in 1938 and grew up in the small town of Warsaw, Indiana. Already in high school he made a name for himself as a runner, and during his college days at the University of Southern California he became a nationally ranked track star. He placed sixth in the 10,000 meters race at the 1960 Rome Olympics and established a U.S. record at that distance.

Truex was diagnosed with Parkinson's disease around 1977,

when he was making a successful career as a lawyer and was married with a young family. He himself attributed his illness to a bout of heat stroke he had suffered during a competition in Mexico City. His disease course was fairly typical, starting with tremor and rigidity in the limbs and progressing over a decade or so to a severe disability in which he had trouble getting out of bed or rising from a chair.

It so happened that Truex's former track coach knew Robert Iacono's father, so in 1988 Truex consulted with Iacono, who was at that time practicing in Tucson, Arizona. Because the adrenal medulla transplants were then very much in the news, Truex wanted to have that operation. Iacono recommended against it, suggesting instead that he have a fetal-cell transplant. Iacono tried to arrange for him to receive one in Sweden or in England, but without success. He also contacted Eugene Redmond's group at Yale, but they were looking for volunteers who lived nearby. Then, toward the end of 1988, the news broke about our first transplant operation on Don Nelson, and a few weeks later the Yale group performed their first transplant, too. Feeling that the procedure had now been legitimized, Iacono resolved to perform the transplant himself. Realizing, however, that anti-abortion attitudes and the fetal-tissue funding moratorium would impede his efforts to perform the operation in the United States, he cast about for an overseas site where he might be able to do it unhindered.

During this period Iacono spent some time in Japan and tried to arrange to perform the transplant there, but his Japanese colleagues were too leery of the potential media reaction to help him. While in Japan, however, Iacono met a Chinese doctor who worked at a cancer hospital in Zhengzhou, a city in central China. This doctor told Iacono that he could perform the transplant there and that fetal tissue would be readily available. Iacono resolved to take Truex to Zhengzhou.

In its geographical audacity, Iacono's plan resembled that of Bill Langston, who took his two California patients all the way to Sweden for transplants. But Langston was taking his patients to a world-famous research group who had done pioneering animal studies on fetal-cell transplants, and who had already performed several transplants on human patients. Iacono, on the other hand, planned to take Truex to a hospital whose staff was utterly unfamiliar with this kind of operation. And Iacono himself had never performed a fetal-cell transplant—or even witnessed one, so far as I know.

Iacono knew that the Zhengzhou hospital lacked the stereotactic and scanning facilities that would allow him to place the transplanted tissue at the appropriate sites in Truex's brain. He therefore devised a scheme to make it possible to perform the transplants at Zhengzhou without the aid of any sophisticated equipment. In April 1989, Iacono operated on Truex at the Veterans Administration hospital in Tucson. There he performed a procedure called a *thalamotomy,* which involves destroying part of a brain structure near the striatum known as the thalamus. This operation can provide significant relief for Parkinson's disease patients—it is particularly effective for the relief of tremor, which was one of Truex's symptoms.

As Iacono told us recently, he mainly performed the thalamotomy as a "cover" for another procedure he carried out on Truex's brain that day—one that would not have been approved if its nature had been known. He implanted three plastic tubes under stereotactic control: two of the tubes led from the surface of Truex's cerebral cortex down to the striatum on each side—to the caudate nucleus on the left side and to the putamen on the right side. The third tube led down to the left lateral ventricle, which is one of the two large fluid-filled spaces that lie within the body of the left and right cerebral hemispheres. He then closed the scalp incision. These three tubes would be the delivery chutes for the tissue he planned to transplant in China.

Iacono recollects a conversation he had with Truex's wife before the two men set out for China. "I said, 'You know, Kay, I may not be able to bring him back.' And she said, 'Well, Bob, you've got to try desperate—' and I said 'Kay, I may not be able to bring him

back even in a box.' She said, 'Bob, please try.' So this wasn't tid-
dlywinks. And I'm no Texas chainsaw murderer, I'm a very conservative neurosurgeon."

With the tubes in place in Truex's brain, the two men set out on
their marathon journey. From Los Angeles they flew to San Francisco. From there they took a China Airlines flight for Shanghai,
but the flight was diverted to an airport in Japan after the aircraft
ran dangerously low on fuel. After a long delay they reached Shanghai, where they embarked on a train for Zhengzhou. Iacono had
never been to mainland China before, and the primitive conditions
soon impressed themselves on him: the train, hauled by a steam
locomotive, took twenty-two hours, pulling off onto spur lines
every time a freight train needed to use the main track.

The long journey and the recent surgery took their toll on Truex.
When the train finally stopped at Zhengzhou station he was not
even able to take the few steps from the train to the waiting car,
and no wheelchair was available. Iacono had to lift him into the
back of a bicycle rickshaw to move him the short distance. Iacono
was further dismayed when they reached the hospital: he found that
it lacked all modern amenities, including heat and hot water. And
in the laboratories the microscopes had no built-in light sources;
rather, they had mirrors that were aimed at a window or a lamp,
like children's toy microscopes here.

Iacono tells us that, before operating on Truex, he needed to
learn how to dissect the fetal tissue. So he said to the Chinese staff,
"I need some things to dissect, guys, because I haven't had any
practice in my country. I need some embryos, some fetuses, bring
me some stuff." A couple of hours later they did. "Where did you
find that?" Iacono asked. "Oh, in the dustbin," was the reply.

A few days later, Iacono put Truex on immunosuppressive drugs
and began the transplant procedure. On the first day of the procedure, tissue from a freshly aborted fetus was brought to him. Iacono
estimated that the fetus was sixteen weeks old, which is considerably older than the optimal age for transplantation. Iacono dissected out the fetal tissue he needed and cut it into fragments. Then,

after reopening Truex's scalp incision, he implanted the fragments in the right caudate nucleus by pushing them down one of the plastic tubes. He then removed that tube from Truex's brain.

A couple of days later Iacono took tissue from another sixteen-week-old fetus and implanted it into Truex's left putamen. This time he did not simply push the fragments down the tube; he first loaded the fragments into a metal coil resembling a long narrow spring. He pushed the entire assembly down the plastic tube into the putamen with the aid of a stiff wire; then he pulled the tube out of the brain while keeping the wire in place to prevent the metal coil from moving back up with the tubing. After he removed the wire, the metal coil, with the tissue fragments within it, was left permanently lodged in the putamen.

Finally, Iacono implanted tissue from a third fetus into Truex's brain. This fetus was five to six weeks old; that is, much younger than the previous two; younger, in fact, than the tissue we and most other researchers have used for transplantation. For the implantation of this third sample, Iacono took a radically different approach from the previous two transplants. He broke up the tissue into very small pieces or single cells and injected the resulting suspension into the left lateral ventricle via the plastic catheter whose tip had been placed there. Having completed the injection, he left the catheter in place and sewed up Truex's scalp.

What Iacono's motivation for doing the third transplant in this way may have been, I do not know, but it was fraught with hazard. The brain's ventricular system is a system of cavities that communicate with one another and with the spaces outside the brain and within the spinal cord; the cerebrospinal fluid circulates steadily through this system. Thus, cells introduced at one point may be carried to other locations within the system, and one has essentially no control over where they may end up. Even though the portion of the ventricular system where Iacono injected the cell suspension was near the caudate nucleus, there was no reason to think the injected cells would remain there, and as was discovered much later, they did not.

Perhaps the most unusual aspect of this whole operation, how-ever, had to do with the precise nature of the fetal tissue that Iacono transplanted into Truex's brain. "I never took dopamine cells for my grafts," he told us recently. "That just shows you my contempt for the dopamine hypothesis."

Iacono had come to believe that the loss of dopamine input to the striatum played only a minor role in Parkinson's disease, and that the loss of other neurotransmitters, including serotonin, was more important. Therefore, in dissecting the fetal brain for trans-plantation, he took tissue from a strip of brain near the midline, extending from the septum, which is near the front of the brain, back to the medulla, where the brain joins the spinal cord. Within this extended zone are found cells that use serotonin and a variety of other transmitters.

Having completed the three transplants, Iacono and his patient made the long return trip to the United States. This time, they flew from Zhengzhou to Shanghai aboard a Russian-built aircraft which Iacono remembers chiefly on account of its tires—they were worn down to the Dacron. A few weeks later that same aircraft crashed, killing all passengers and crew. Truex and Iacono got back to the States safely, however, and Truex paid Iacono about $30,000 for the transplant procedure and associated costs, according to news reports.

I do not know the details of Truex's postoperative course, but a year later Iacono was reporting that he was much improved. Truex was walking, swimming, and climbing stairs, he said. He added one detail that resonated with my own experience of transplant recipi-ents: Truex borrowed Iacono's four-wheel drive vehicle and went driving around the Mojave Desert in it, luckily without incident.

How much of this apparent improvement was due to the fetal-cell transplant is hard to say. Truex had had a thalamotomy, after all, and after the transplant Truex's doctors made changes in his drug regimen, including the addition of a new drug, Deprenyl. This drug blocks the breakdown of dopamine and thus increases its con-centration in the striatum. Since both thalamotomy and Deprenyl are known to improve the condition of some patients with Parkinson's disease, one has to be cautious in attributing Truex's apparent improvement to the transplant.

Both Truex and Iacono felt that the transplant had been beneficial, however, and Iacono subsequently took several other patients to China to undergo similar operations. In addition he performed some transplants on Chinese citizens. With these later patients, Iacono carried out the transplants at a better-equipped hospital in Shanghai.

During the second year after his transplant, Truex visited several neurologists around the country. Apparently, he was motivated more by the wish to "show off" his transplant than by a need for additional neurological care. At around this time Truex and his wife moved to Boston, and one of the neurologists he called on was Raymon Durso, who runs a Parkinson's disease clinic at the Veterans Administration hospital in that city. (Durso is also affiliated with Boston University Medical School.) Durso was not able to make any objective assessment of whether the transplant had benefited Truex or not, since he had not seen Truex before the surgery. But he did confirm that Truex was able to get out of his chair unaided, something that Truex's wife said was not possible for him previously.

Truex visited Durso on two or three occasions, the last time being in March 1991, nearly two years after his transplant. On that occasion Durso noted some edema in Truex's legs and a raised level of urea in his blood, suggestive of mild kidney failure. This could well have been a side effect of the ongoing treatment with immunosuppressive drugs. In other respects Truex seemed reasonably well.

One Sunday morning just two weeks after this final visit, however, Truex woke up feeling very tired. Later, he fell back asleep, and his wife noticed "funny halted breathing." She called an ambulance, but by the time paramedics arrived he had stopped breathing altogether. He was brought to the New England Medical Center, which is Tufts University's teaching hospital, located in Boston's Chinatown. There, he was pronounced dead.

The New England Medical Center staff had no interest in conducting an autopsy, because Truex had died at home of what seemed to be natural causes. But Truex's wife called Iacono to tell him what had happened, and Iacono very much wanted to have Truex's brain examined so that the condition of the transplanted cells could be determined. He therefore asked Durso to arrange for a brain autopsy.

Durso tried to contact several neuropathologists at the V.A. hospital who might have been able to conduct the autopsy. Perhaps because it was Sunday evening, he had no luck. Then he paged several other neuropathologists in the Boston area, but again drew a blank. Finally he called Rebecca Folkerth, a young neuropathologist who was on the New England Medical Center staff, but who on that day was on call at a different hospital.

"I was one of the foolish people who leave their beepers on all the time," said Folkerth recently. "So I said, 'OK, I'll come and do this autopsy.' I went there to remove the brain, and as I got there I received a call from Dr. Iacono, who told me the patient's history, and he said, 'Can I ask you to take some of the fresh tissue and freeze it?' So I took out the brain and I cut it up fresh, which is not the usual thing—we usually place the entire brain in formalin to harden for a couple of weeks before we examine it."

Iacono's request to have the brain sliced in a fresh condition was motivated by a desire to carry out a procedure called *immunohistochemistry*. This is a method of detecting the presence of specific molecules, such as neurotransmitters or the enzymes that produce them, in slices of tissue by exposing the tissue to antibodies against those molecules. Thus immunohistochemistry can help identify what kinds of nerve cells have survived in the transplants— whether dopamine cells, which would be the usual kind of cell in a transplant of tissue from the substantia nigra, or the serotonin-containing or other cells that Iacono had transplanted.

Because the condition of the transmitter and enzyme molecules can be degraded by long fixation in formalin, Iacono wanted the tissue to be sliced up right away and stored in a freezer. Thus, instead of simply placing the entire brain in formalin, Folkerth settled in for a longer procedure, even though it was now well after midnight. She took a long, broad-bladed knife resembling a spatula with sharpened edges. With it, she began cutting the brain into a stack of sections about half an inch thick.

She had hardly started the slicing procedure, however, when she got the shock of her life. She saw, growing deep within the brain—hair! And then she saw cartilage. These and other, less easily recognizable tissues formed nodules that were scattered through the ventricles of Truex's brain, in places filling the ventricular spaces completely.

"It was the strangest thing I'd ever seen," says Folkerth. "It was creepy. It was the middle of the night and I was alone in the autopsy room, and I'm thinking, 'What the hell is this? This is some mistake.' "

Collecting herself, Folkerth cast around for some explanation of how these tissues could have grown inside Truex's brain. Her first thought was that this was a rare kind of tumor called a teratoma, which can contain a wide variety of tissues including hair, cartilage, bone, and even teeth. A teratoma most often forms in the ovary, and it derives from reproductive cells that somehow maintain their ability to form multiple cell types. It is almost as if the tumor is trying to form a human fetus, but with absolutely no sense of where everything is supposed to go. It looked as if the same thing was happening in Truex's brain.

Iacono had not mentioned any tumor, though, only the fetal-cell transplant. And when Folkerth examined the slices more closely, she saw evidence of the transplant procedure Iacono had carried out. In the left putamen she saw the metal coil Iacono had implanted. In the right caudate nucleus she saw an area of scarring that seemed to correspond to the location of the transplant on that side. She also saw the plastic tube that ran down from the brain surface to the left lateral ventricle. She began to wonder if the strange nodules scattered through the ventricular system were somehow related to the tissue injection that Iacono had made through that tube.

Folkerth also began to get an idea of how Truex had died. The brainstem, which controls breathing and other vital functions, showed signs of compression, either from the growth of the nodules or from a blockage of fluid flow within the ventricular system caused by their presence. Folkerth could not rule out other possible causes of death, however, such as a heart attack, because she did not have permission to autopsy the remainder of Truex's body, and her expertise was limited to the brain. In fact, Iacono still maintains that a heart attack was the likely cause of death.

After photographing the brain slices and taking samples for immunohistochemistry as well as for conventional microscopic study, Folkerth was finally able to go home and get some well-deserved rest. But the bizarre case stayed very much on her mind, and over the next few months she spent a lot of time analyzing the tissue samples that she had taken. Under the microscope, she could see no surviving nerve cells at the implantation sites in the left and right striatum, only scar tissue. Within the scar tissue were inflammatory cells—a possible sign that Truex's immune system had mounted an attack on the fetal transplants in spite of the immunosuppressant drugs he was taking.

Folkerth was also unable to detect transplanted nerve cells with the immunohistochemical technique. However, she used an antibody against tyrosine hydroxylase, an enzyme found in dopamine cells. This was probably because she thought that Iacono had transplanted dopamine cells. Cells containing serotonin or most other transmitters would not be detected with such antibodies. At any event, it did not seem that any transplanted neurons, if they survived the operation at all, were still present at the time of Truex's death.

When she examined the brainstem under the microscope, Folkerth found evidence that it had been subjected to chronic compression, strengthening her suspicion that Truex's death had been caused by damage to the respiratory centers there. "I think he had gradual changes in his brainstem that couldn't be compensated for any longer," Folkerth says.

As for the nodules themselves, Folkerth confirmed under the microscope that they contained a variety of tissue types—hair, cartilage, skin, and even bone—that had no business being in the brain. It seemed to her that the nodular tissue must have derived from the cells that Iacono had injected into the left lateral ventricle. When she learned how Iacono had taken tissue from a very young fetus, from a very extensive area of the brain, and with the aid of a poor-quality microscope, her suspicions gelled. Probably, Iacono had inadvertently included tissue that did not belong to the fetal brain at all, but to nearby fetal structures that normally develop into skin, cartilage, and so on. Those cells, once transplanted into Truex's brain, had apparently drifted through the ventricular system, taken root at many different sites, and then followed their own predetermined developmental pathway.

Folkerth discussed her findings with Raymon Durso, and also with Iacono. Both Folkerth and Durso thought that the findings were so unusual—and potentially relevant to fetal-tissue transplants in general—that they should be written up for publication. But Iacono seemed to be averse to the idea of publishing it. This was a major change of heart on his part, because he had been the driving force behind having Truex's brain examined in the first place. But it was an understandable change, perhaps, given that the autopsy findings suggested to Folkerth and Durso not only that the grafts had failed but also that they contributed to Truex's death.

An autopsy study of this kind would normally be written up as a collaborative paper authored by the pathologist (Folkerth), the neurologist who had been seeing Truex (Durso), and the surgeon who had performed the transplant (Iacono). So when Iacono lost interest in seeing any publication come out of the study, Folkerth and Durso were in a bit of a quandary. They let the matter slide for quite a while, but eventually they decided to write up their findings and submit them for publication with just themselves as authors. They submitted their manuscript to the *New England Journal of Medicine*. Their reason for doing so was that the *New England Journal* had recently published several quite positive papers (including our own) on the topic of fetal-cell transplants for Parkinson's disease. It seemed to Folkerth and Durso that their paper, which documented an unanticipated and apparently fatal complication of the procedure, was a cautionary tale that would be of great interest to the journal and its readers.

As is standard practice, the editor of the *New England Journal* sent the manuscript out to a couple of scientists for peer review. One of these scientists was myself. When I read the manuscript I was shocked by the story it told. I did not think it should be published by the *New England Journal*, however. That journal prides itself on publishing papers of broad medical and scientific importance. This manuscript, on the other hand, described a therapeutic misadventure. I was concerned that if it were published by the *New England Journal* some readers might misinterpret it as demonstrating a fundamental flaw in the fetal-cell transplant procedure. This in turn could bring the procedure into wide disrepute. In my review letter to the editor, I suggested that the manuscript be declined but that the authors be encouraged to submit it to a specialty journal.

I don't know who the other reviewer was or what his or her comments were, but they must have been somewhat in line with mine, because the editor of the *New England Journal* did in fact reject Folkerth and Durso's manuscript. After that, the authors let their manuscript sit in a desk drawer for a while, but in 1995 they resubmitted it to another journal, *Neurology*. This journal did accept the manuscript, and it finally appeared the following year, five years after Truex's death.

In the paper, Folkerth and Durso refer to Iacono obliquely, so that a casual reader would not know who had done the surgery on Truex. To people in the field, on the other hand, his identity was obvious. I do not know whether the paper's publication caused any problems for him, but by the time of publication he had given up doing fetal-cell transplants altogether and was concentrating on another surgical procedure called pallidotomy (see chapter 13). He tells us that he has now done two thousand of these operations at Loma Linda University Medical Center. Aside from his belief that pallidotomy is superior to fetal-cell transplantation for Parkinson's disease, Iacono has also become strongly religious and "pro-life" since his fetal-transplant days. His waiting room has a painting of a surgeon operating, with Jesus standing next to him and guiding his scalpel.

Aside from a mention of two patients in a brief 1992 report, and some Chinese-language papers, Iacono did not publish any results for the other patients on whom he performed fetal-cell transplants. He says that none of the other patients suffered adverse consequences.

One person who reacted strongly to the publication of Folkerth and Durso's paper was Jeffrey Kordower, a neuroscientist at Rush-Presbyterian–St. Luke's Medical Center in Chicago. Kordower was a member of a group led by Thomas Freeman of the University of South Florida in Tampa that had started doing fetal cell transplants for Parkinson's disease in 1993. Kordower wrote a letter that was published in a later issue of *Neurology*, in which he was sharply critical of what had happened to Truex.

As far as I was concerned, Max Truex died because of a medical misadventure that should never have happened. It did not reflect on the basic value or safety of the fetal-transplant technique at all. Yet we knew there were intrinsic risks to neurosurgery and that it was just a matter of time before one of our own patients would have a catastrophic surgical complication.

CHAPTER 12

A Tragedy and a New Beginning

In January 1994 one of our fetal-transplant patients died as a result of a brain hemorrhage he suffered during the transplant operation. This tragedy, though not totally unexpected, caused us to reevaluate our program and make important changes in the way we carried out the transplants.

The patient—our seventeenth—was a fifty-four-year-old man named "Gerald Walster." The operation began in the usual way. As was our custom, we performed the transplant on the left side of the brain first. We do it this way because the left hemisphere is generally the dominant hemisphere, and we want to make sure that we encounter no problems on that side before proceeding to the nondominant right hemisphere.

After we had completed eight needle passes into the left putamen, we began to repeat the procedure on the right side. The first five passes on the right side went uneventfully. As we withdrew the needle after the sixth pass, however, we noticed that the outside of the needle had a thin coating of blood. In addition, a small amount of blood oozed from the site where the needle had entered the brain.

Usually, the needle comes out of the brain completely "clean," so we knew that there had been a hemorrhage somewhere along the needle track. However, the oozing of blood from the needle puncture stopped quickly, so we thought the bleed was probably small in quantity and located superficially. We therefore decided to continue the operation and made the seventh and eighth needle passes without incident.

As we wound down the operation, we noted that Walster had some weakness in his left hand. Because of the crossed connections between the brain and the body, that meant something was wrong on the right side of his brain. Then, while he was still in the operating room, his entire left side became weak, and he also began to have difficulty answering questions we put to him.

We took Walster down to the basement for a CAT scan, and on the scan we saw that there was a significant blood clot in the white matter—the fiber bundles—under the cerebral cortex but above the striatum. By the time we saw the scan, Walster was barely responding to questions or to physical stimuli. Undoubtedly, the clot in the white matter was putting pressure on the undamaged brain tissue in the vicinity and thus preventing blood from flowing through it.

Bob Breeze and I had dealt with this problem before. Basically, brain hemorrhages can be treated surgically by evacuating the blood clot, or medically by giving drugs that counteract brain swelling. Surgical treatment of a hemorrhage is risky because it can end up making things worse. But if the bleeding continues, it compresses the brain, causing more damage. Within a few hours after the completion of the original surgery Bob Breeze took Walster back to the operating room. Bob reopened the scalp wound, removed the bone flap, and suctioned out the clotted blood.

Unfortunately, Walster did not improve significantly as a result of the second operation: he remained paralyzed on his left side. He continued in this state for several weeks. During that time he had a series of crises including a urinary tract infection and seizures. On occasion, he seemed to be making progress; at other times he was much less unresponsive, except for withdrawal of his limbs in response to painful stimuli. Eventually he developed pneumonia.

About a month after the transplant operation he suffered a cardio-respiratory arrest. He was resuscitated, but an MRI scan the next day showed that his brain had a so-called ground glass appearance. This was a sign that the brain had been irreversibly damaged by the lack of oxygen during the arrest. With the family's consent, he was taken off the respirator and died.

This tragic sequence of events was a terrible blow to everyone involved. We had known that making needle passes into the brain carried a risk of hemorrhage, and that such a hemorrhage could sometimes be fatal. Our estimate, based on the long experience of Breeze and of neurosurgeons in general, was that each needle pass carried a 1 in 500 chance of causing a hemorrhage that would lead to a serious stroke or even death. Since we were making sixteen needle passes into each patient, we had estimated that 3 percent of our patients would have this complication. We have now performed sixty-one transplant operations, and Walster has had the only fatal complication.

We had explained to all our patients that they had a chance of experiencing a hemorrhage, and that such a hemorrhage could be fatal. But to know intellectually that such a risk existed, and to see it happen to a particular patient, were two very different things. For the patient who actually suffers a hemorrhage, it is not a 2 percent risk—it is a 100 percent reality.

After this tragedy, we suspended the transplant operations for several months until Bob Breeze came up with a novel solution for reducing the risk of the operation by reducing the number of needle passes we were making into the brain. Bob's idea was to change the direction from which we approached the striatum. The putamen—the portion of the striatum that was our target for the transplants—is a roughly olive-shaped structure, with its long axis oriented front to back in the brain. With the standard vertical approach that we had used up until then, it took eight passes, spaced at 4-millimeter intervals, to cover the front-to-back extent of the putamen on one side of the brain. Bob's idea was to make horizontal penetrations, along the long axis of the putamen. With this approach, only two passes, one above the other, would be required to cover the putamen on one side of the brain, and therefore a total of only four passes for a bilateral transplant: one-quarter the number of passes we had been making up until then.

To make horizontal penetrations meant starting at the front of the head instead of the top of the head. Bob's plan was to make an incision in the skin of the forehead, an inch or so above the eyes, drill small holes in the skull there, and run the needle back to the putamen and through its long axis to its most posterior margin. Although the needle passes would have to be a little longer than with the vertical approach, the total length of the combined passes would be much shorter.

A second change concerned the needles we used for the transplants. We conferred with Trent Wells, the maker of stereotactic equipment who had been helping us from the beginning of our program, and he designed and made a new needle assembly. The entire needle was about a foot long, and it had three different widths, rather like a three-stage rocket. The four or five inches farthest from the tip was the thickest part of the needle; it was 1.5 millimeters in diameter. This portion never entered the brain, it simply spanned the distance between the stereotactic apparatus and the patient's head, so it was made thick enough to be completely resistant to bending. The middle portion of the needle was 1.0 millimeters in diameter: this portion did enter the brain but not the striatum. The five centimeters of the needle closest to the tip was only 0.6 mm in diameter. To further reduce the chance of injury to the brain, a rounded stylet was in the bore of the needle while it passed into brain. Once in place, the inner stylet was removed and replaced with a very narrow and extremely thin-walled needle that carried the "noodle" of fetal tissue. As before, we extruded the tissue into the striatum while we were withdrawing the entire needle assembly.

The first patients in whom we used the new procedures were a fifty-nine-year-old woman from Minnesota named Bryce TenBroek and a fifty-six-year-old man from Ohio named Dan Stewart. I always think of Bryce and Dan together, because we operated on them within three weeks of each other in May 1994, and they and their families got to know each other very well during the process of evaluation prior to surgery. Afterward, they asked to have their follow-up visits scheduled at the same time so they could continue to see each other.

Bryce was (and still is) a trim, pleasant woman, a mother of three who is much involved in the social life of a large extended family. Before becoming incapacitated by her illness she was a keen sportswoman and was particularly fond of tennis, which she played several times a week. She also worked in a clothing store.

Bryce's illness began in 1981 with a tremor and tingling in her left leg. The symptoms quickly spread to her left arm. It took four years of visits to doctors and chiropractors before her condition was correctly diagnosed. With the help of L-dopa and other drugs, she remained active until the early 1990s, when the right side of her body became involved. Then her physical abilities declined rapidly and she began to experience dyskinesias and profuse sweating, both of which caused major problems for her at work.

In early 1992 Bryce's sister sent her a clipping of a story from the *New York Times* about our transplant program. Titled "Patients Paying to be Subjects in Brain Study," the story focused on the question of whether it was ethically acceptable to charge patients for transplant operations while the procedure was still to some degree experimental. Eugene Redmond, who had private funding to do transplants at Yale University, told the *Times* he would rather stop his program than charge patients for the operations. Bill Langston in California, who had also raised private money to take his MPTP patients to Sweden, expressed the same view.

Bryce, however, paid more attention to the views of Chris and Robert Orth, who told the *Times* that Robert was so improved that they would do it again, even if the operation cost far more than it did. Bryce's husband Bill TenBroek called the Orths and got more encouragement to go ahead, so Bryce contacted me and underwent a lengthy evaluation. In early 1993, we accepted her into the program, and she was operated in May of 1994, four months after the operation on Gerald Walster. Her wide circle of relatives and friends contributed to the cost of the surgery.

All went well with Bryce's surgery. As far as she herself was concerned, the main difference was that she did not have to have her hair shaved off, since the incisions were made in the forehead rather than in the scalp at the top of her head. Some time before, she had bought an expensive wig to see her through the postoperative period, but she never had to use it. Also, the reduction in the number of needle passes from sixteen to four speeded up the operation. It took only one hour to do the actual tissue implants.

For several weeks after surgery, Bryce's condition was worse than before. She complained of muscle spasms in her legs that wakened her from sleep. But by three months after surgery she had recovered to about her preoperative condition. Then, in early September, she found she could get out of bed and walk before taking her morning drugs. To Bryce, this was a minor miracle, because for four years previously the expression "crawl out of bed" had been a daily and literal reality for her.

Thus began a period of slow but steady improvement. By a year after the operation we had been able to reduce Bryce's L-dopa by about half, and she was virtually free of dyskinesias and cramping. She could walk normally before taking any drugs first thing in the morning. She had begun to play tennis again and had even done some cross-country skiing. The only downside to her generally remarkable improvement is that she began to have occasional episodes of freezing—something she had not experienced before the surgery. The last time she had arrived at Denver airport she froze up as she was stepping out of the train that conveys passengers between concourses. With her husband's help she was able to exit the train before it carried her off for another loop around the airport.

Bryce wrote a book about her experiences, titled *Playing the Hand*, which she published and distributed herself. I can't resist quoting the final paragraph of the book. She wrote:

> It has been three years since we started on our journey;
> at that time we didn't know where it would lead or
> what it would bring. The operation has given us a
> whole new beginning filled with optimism for the

future! We are so thankful to the doctors for their skill and concern, and to our family and many friends for their unending love and support. We don't know what's ahead of us, but then, I guess, no one does. For now it's a miracle and we thank God for his blessing.

Dan Stewart is an accomplished man with a Ph.D. in ceramics technology. In 1972, at the age of thirty-four, he was surprised when friends told him he was dragging his right foot while walking. The following year he developed a tremor in his right hand. His disease progressed quite slowly, but by the mid-1980s he was incapacitated by a variety of symptoms: slurred speech that even his wife had difficulty understanding, handwriting so poor he couldn't read it himself, and problems with gait and control of his arms that forced him to give up golf and racquetball and his favorite outdoor activity—wilderness hiking.

Unfortunately, things only got worse. He developed severe dyskinesias that interfered with his job—he was president of an industrial ceramics company. Soon after taking his dose of L-dopa he would go into a bout of bizarre posturing in which his legs seemed to wrap around his body. Driving became a nightmare. "His arms and legs were going every which way, and he couldn't keep his feet on the pedals," recalled his wife, Marianne, recently. "He shouldn't have been driving. I was torn between thinking about safety—his and the other people on the road—and the wish to allow him what little independence was left to him. I knew that I would soon need full-time help looking after him, or else he would have to go into a nursing home." Perhaps the final indignity, though, came when his grandson was born in 1992—Dan was unable to hold the baby or interact with him in any meaningful way. That, more than anything, was what led him to seek relief from his condition.

Dan Stewart was interested in the early news reports about fetal-cell transplantation, but he had ethical concerns. He and Marianne belong to the Church of the Nazarene, a conservative Christian denomination, and consider abortion sinful. After discussions with their pastor and others, however, they decided that the use of aborted fetal tissue was acceptable as long as the mother's decision to donate the tissue was separate from her decision to have an abortion. Their decision to undergo a transplant caused some controversy within their close-knit congregation, however; one couple actually left their church over the issue, and a subsequent pastor had more negative views than the one they consulted.

Dan's surgery went without incident, except for one thing. As Bob Breeze was attaching the "halo" before the operation, Dan asked out of the blue: "Has anyone ever died in this kind of surgery?" We had previously explained to Dan that the surgery carried a risk of complications, including fatal ones, and the consent form he signed described that risk. I reminded him of that risk and that one patient had died as a result of the surgery. Dan didn't seem particularly concerned. "I had my sights set, so it didn't make any difference," he said recently. "Engineers just want to know all the details."

During the year after his surgery, the videotaped testing showed a steady improvement. His voice, handwriting, and gait returned to near-normal, and his dyskinesias abated as he began to reduce his L-dopa doses. These changes were very apparent to Dan and Marianne, of course. "By three months, his movements weren't quite as wild," Marianne said recently. "By nine months it was miraculous." Following the tradition established by his predecessors, Dan set out on several long car trips with his wife, in which he did at least half the driving and performed flawlessly. The most touching sign of his recovery, though, came with the birth of his granddaughter, whom he was able to hold and play with like a normal grandparent.

Stewart was able to reduce his L-dopa dose by small increments. By a year after surgery, he was down from about 1,000 milligrams per day to about 150 milligrams per day. Then, in October 1995, at his next visit to his neurologist in Ohio he said, "Why don't I go off L-dopa altogether?" He was the first of our patients who was able to discontinue all Parkinsonian medications. He stopped taking L-dopa and did fine.

Stewart is not cured of his disease, however; that is something we have not yet achieved with any patient. Over the past few years he has experienced a mild worsening of his condition: his handwriting has deteriorated somewhat, he has occasional difficulty with his speech, and he also has some involuntary movements. Finally, he complains of mild cognitive impairment: he says he has difficulty with solving puzzles, doing mental arithmetic, operating gadgets, reading maps, and finding his way around locations not near his home. Unfortunately, cognitive problems occur in many patients with Parkinson's disease; these may be caused by loss of dopamine in other areas of the brain. But in general Stewart is immeasurably better than he was before surgery.

We performed transplants on six other patients in the two years after Bryce TenBroek and Dan Stewart. Then, in May 1996, we performed a repeat transplant on our pioneering patient, Don Nelson, who had been going downhill.

For four or five years after his initial transplant in 1988, Don Nelson did extremely well. Stuart Schneck, who had been Don's neurologist since 1973, was able to cut his L-dopa dose by half. Don's tremor was entirely gone and his rigidity nearly so. He did continue to experience episodes of "freezing," especially when entering elevators, passing through doorways, and the like. But his general state of well-being was well illustrated by an account he wrote himself in early 1992:

> I feel that this past spring and summer were my best in several years. I was very active physically, helping my son to finish his house with framing, putting up drywall, and painting. I built or refinished several pieces of furniture for family, friends, and neighbors. We also completely remodeled two bathrooms. Since my son got married and moved out of the house I have taken over his bedroom and am building a model train layout with Lionel trains that I had packed away since 1951. A few

years ago my hands would not have been able to solder
and work with small wires. This is a cold weather project as I prefer to work in my yard in the summer. This
was all proof that the fetal-cell implant is a success.

I am a model railroad buff myself—I have a large scale-model
train set up permanently in my backyard—so I can attest to the
complexity of Don's layout, which includes hundreds of feet of
track, every kind of locomotive and rolling stock, scale-model vil-
lages, bridges, turntables, and the like, all remotely controlled from
Don's central command post.

Don also devoted a lot of energy to dealing with the media.
Starting a few months after his surgery, he agreed to be interviewed
by reporters—largely, I believe, in the hope of winning support for
our research program. The media were not slow in taking him up
on his offer. Within a couple of years he had been interviewed by
leading newspapers and newsmagazines in the United States, as well
as by several European papers. He also appeared on local and
national TV network news programs, as well as CNN and PBS.

As another sign of his well-being, in 1992 he decided to start a
business again. But this only lasted a few months before he decided
it was too much for him. Instead he concentrated on one of his
favorite hobbies, woodworking; he spent up to fourteen hours a
day in his woodshop making small articles and gifts, which he distributed to family and friends.

We did some objective tests to monitor Don's continuing
improvement. In 1991 we sent him back to UCLA for another PET
scan—his third. This scan, done three years after transplant,
showed increased uptake of fluorodopa in his right striatum, compared with the scan that had been performed nine months after his
transplant. These results, when published in the *New England Journal of Medicine* in 1992, provided the first evidence that transplants
continue to develop over many years.

To study his improvement more objectively, in January of 1992
I set up the computer testing equipment in Don's home once more.

For four weeks he tested himself daily, just as he had done before and after the transplant. The resulting data showed that his hand movements maintained the improvement they had shown during the first year after transplant, and his walking speed before taking his morning medications had improved beyond the level it had reached during the first year. His drug doses remained at about half the pre-transplant levels.

Beginning around the latter part of 1992, however, Don noted some deterioration in his condition. He began to be bothered by dyskinesias affecting his head and neck, and his general energy level decreased somewhat. Stuart Schneck had to gradually increase his L-dopa dose again, but even so Don's condition continued to worsen, though gradually. By 1994, his walking speed had decreased noticeably and he was depending on his walker and crutches more than he had to up until then.

This gradual worsening resembled the usual course of Parkinson's disease. It therefore seemed likely that it was caused not by any problem with the transplant itself, but by the continuing loss of Don's own dopamine neurons from his substantia nigra. We had only made a transplant into one side of Don's brain, and then only injected cells from one embryo. Very likely the transplant had been able to compensate for the moderate level of dopamine cell loss as it affected Don in 1988, but could not cope well with the increasing losses that took place over the following four or five years.

Our original decision to perform a transplant into one side of Don's brain had been made out of caution; we simply didn't know what the effects might be so kept the risks as low as possible by confining the surgery to one side of the brain. In fact, when it became obvious that a one-sided transplant led primarily to benefit only the side opposite the transplant, Don brought up the issue of a second transplant.

It seemed reasonable to believe that a second transplant would help Don. After all, he had had Parkinson's disease for twenty-five years, so his left striatum was probably running on empty, as far as dopamine was concerned. Certainly his PET scan showed a severe decrease in fluorodopa uptake on that side. There was only one issue that really concerned me, and that had to do with Don's immune system.

As I described in chapter 6, we had a bad experience when we made two successive transplants into one of our MPTP-treated monkeys. The first transplant, into the right striatum, alleviated the monkey's symptoms on the left side of its body, as we had hoped. But when we made a transplant into its left striatum seven months later, the monkey's immune system mounted an attack on *both* the transplants, and the monkey's condition deteriorated within one week to its preoperative state. Evidently, the second transplant had reminded the monkey's immune system that there were foreign cells lodged in its brain, and the result was a massive immune response.

We certainly didn't want anything like that to happen with Don. We had strong reason to believe that the initial transplant was still functioning. If that transplant were to be knocked out by an immune response triggered by a second operation, Don might be thrown into a Parkinsonian state more profound than anything he had experienced to date. I consulted with several immunologists about whether we should proceed with a second surgery or not.

There was one aspect of the monkey episode that was potentially relevant. The two monkey fetuses that had supplied tissue for those two transplants were related: they had the same father but different mothers. The immunological literature indicated that a second challenge with the same or related tissue produced a rapid and intense immunologic response (the "amnestic," or "remembered" response). With Don, any second transplant would come from unrelated donors, so that the risk of an immune response was less.

After a fair amount of consultation and soul-searching, we decided to go ahead and perform a transplant into Don's left striatum. To minimize the risk of an immune response, however, we decided to give him immunosuppressant drugs for a year after the transplant. Don participated in these discussions and agreed with our plans. These plans were approved by the IRB, and Don signed a special consent form spelling out the risk.

We carried out the second transplant in May 1996, seven and a half years after his first. Don was pleased with the improvement in our surgical prowess over the intervening years, which allowed us to shave about six hours off the duration of his operation. The operation went off without incident, and Don did not suffer any worsening of his condition such as the MPTP-treated monkey had experienced, either soon after the operation or a year later, when we discontinued the immunosuppressant drugs. He also failed to experience any great improvement after the operation, however, which was disappointing.

Part of the reason why Don didn't improve significantly may have had to do with several other health problems he suffered during the postoperative period. In December 1997 he got a splinter in his hand while woodworking. Carolyn got the splinter out and treated the skin with peroxide, but nevertheless Don developed a serious staphylococcal infection that required hospitalization. The infection spread to his lungs and then to his hip. By the time he was over it, he had lost forty pounds. Then he developed problems with his back that required surgery. It was 1999 before he was getting back into reasonable shape.

Recently, Don told me he feels he significantly benefited from the second transplant, even though the improvement was delayed by his other illnesses. I hope that's true, but without the kind of objective testing we did earlier, I really can't verify his assessment. He is clearly much better than he was prior to his initial transplant—he does not have to crawl around the house on hands and knees as he did back then—but equally clearly he is someone who is still battling Parkinson's disease.

The reason that we haven't done the computer testing with Don this time is that we have been absorbed with a larger project—the federally funded double-blind study that finally got rolling in 1995. To explain how that project got started, we have to step backwards a few years, to 1992.

CHAPTER 13

Double-Blind

Although the attempt to roll back the fetal-tissue funding ban in early 1992 was a failure, everything changed later that year. At the start of the 1992 presidential campaign, it seemed likely that George Bush would be reelected: he had all the advantages of incumbency, and as commander in chief of the armed forces he had a resounding victory in the Gulf War under his belt. So good were his prospects that Democrats were given little chance of success. Then a wild card appeared in the form of Reform Party candidate Ross Perot. Through the primaries, Bill Clinton gained strength. On November 3, after a tumultuous campaign that included three-way debates, Clinton won an electoral landslide of 362 votes.

The ban was an issue during the presidential campaign. Bush, of course, supported it, while Clinton wanted to end it. As soon as I heard that Clinton had been elected, I assumed that the regulations would change. Within a week of the election, Breeze and I met and decided to submit an application for an NIH grant to do a clinical trial of fetal-cell transplants.

Our plans were complicated by the fact that I was in the middle of a sabbatical year. Nancy and I had passed the first part of the year in France—I spent the time at a lab near Paris where I learned molecular genetic techniques that I hoped to apply to the treatment of Parkinson's disease. After that, we returned to Denver for several months so that Bob Breeze and I could do some transplant operations. But on November 26—Thanksgiving Day—we were scheduled to leave by road for the East Coast. We were moving to Baltimore for the remainder of my sabbatical year, so that I could work at the laboratory of George Uhl at Johns Hopkins Medical School.

Winter can come early in Denver—snow often falls in September. Six days before our departure, it started snowing. At the same time, I got a blizzard of requests for interviews, because the media had been alerted to the forthcoming publication of a report in the *New England Journal of Medicine*, in which we described the results of our first seven fetal-cell transplants.

In the *NEJM* paper we reported several novel findings. First, that most patients had an increase in dyskinetic movements after their transplant, which made it necessary to cut their drug doses by about 40 percent in the first year. Second, that transplant effects seem to develop over a period of years. Don Nelson, for example, was better four years after transplant than he had been at fifteen months. A fluorodopa PET scan of his brain, done by John Mazziotta at UCLA, showed more uptake three years after transplant than nine months after. Others later confirmed these results.

Interviews continued for the next week, even as we loaded two cars for our journey. On November 23, a second and bigger snowstorm struck. One of our cats, named Orange Pop, holed up somewhere and was not to be found. We searched for the next three days, but with no luck. On Thanksgiving morning we headed east with me in my Toyota truck and Nancy and our cat Celeste in her Honda, both vehicles packed solid.

Because the blizzard had moved directly east, we decided against I-70 and headed northeast to I-80 to cross Nebraska. There the snow had ended, but signs of the storm's severity were everywhere: abandoned trucks lay jackknifed or overturned along the highway. Nancy and I stayed in touch via CB radios that we had bought for the occasion, but long periods went by without conversation, and I used the time to think about our plans.

When Bob and I had discussed our grant proposal, we had come to realize that we needed to think bigger than we had in the past. We had improved our techniques to what seemed like a reliable level, we were performing about five transplant procedures a year, and we were gradually accumulating data on the long-term effects of the transplants on the men and women who received them. We were satisfied that our methods were refined enough for a larger study that would test the efficacy of the transplant procedure in a more effective way. A problem with embarking on such a study, however, was that the University of Colorado lacked a large Parkinson's disease clinic: we did not have a big enough patient base from which we could recruit volunteers for the study. We therefore needed to bring in other groups.

As we traveled eastward, newspapers and the nightly television news carried reports of our *NEJM* article. There were actually two other similar reports in the same issue: one from Olle Lindvall and Anders Björklund's group in Lund, and the other from Gene Redmond and Dennis Spencer's group at Yale. In accompanying editorials, the *NEJM* raised the ethical and scientific issues involved in fetal-cell research. The scientific overview was written by Stanley Fahn, an internationally recognized expert on Parkinson's disease based at Columbia-Presbyterian Medical Center in New York. I had known Stan for years, and Bob and I had discussed collaborating with his group. On Monday November 30, we were on the Pennsylvania Turnpike, the final leg to Baltimore, when I called Bob on my cell phone to ask if he thought we should get in touch with Fahn. He said yes.

There were several reasons why I thought that a collaboration with Fahn could be key to a successful large-scale study. I had known Stan for many years. He is clear thinking and open to new ideas. At a purely practical level, Fahn's clinical research unit has access to a very large patient base, so that finding suitable volunteers for the study would be relatively easy. Also, Fahn's clinical experience would be invaluable.

The day after we arrived in Baltimore, we got word that our cat Orange Pop had been found. He arrived by air the day after. Our household was intact!

Shortly after getting to Uhl's lab at Johns Hopkins, I called Fahn. He was immediately interested in the transplant project. We agreed to get together in New York later in December, on the twenty-second. On the seventeenth, I called to finalize our meeting. It was during that conversation that Stan made the remarkable suggestion that the study should include placebo surgery. Since sham operations are rarely used in surgical research and had never been done in neurosurgery, it was a bold proposal. I called Bob to see what he thought: he liked the idea. With two phone calls, we had our plan.

In a double-blind study, neither the patients nor the researchers studying them are allowed to know which patients got the experimental treatment and which the control treatment. This way of conducting a study ensures that neither the patients' nor the doctors' personal beliefs about the efficacy of the experimental treatment influence the study's final outcome.

Double-blind trials have long been standard practice in research involving drug treatments and vaccine trials. They often generate a certain amount of controversy. In drug studies, patients and ethicists often complain that placebo patients are being denied a potentially lifesaving medication. Sometimes the facts prove otherwise. In a test of a heart medication called encainide, for example, patients who received the real drug proved much more likely to die than patients taking the placebo medication, demonstrating that a new treatment can make people worse. In a recent trial of an AIDS vaccine, one-third of the volunteers were given a dummy vaccine. All the volunteers were told that they might receive a dummy injection, and that the vaccine itself was of unknown value. All the volunteers were counseled on the use of safe sex techniques. Yet there were critics who suggested that the mere fact of receiving an injection of something, and the knowledge that it had a two-thirds chance of being a vaccine, was liable to cause some volunteers to engage in unsafe sex and thus possibly to acquire HIV infection.

Even with these ethical concerns, double-blind drug trials are strongly preferred because they are the only way to obtain truly objective data about a drug's efficacy. Nearly every new drug must be tested by a double-blind trial before approval by the FDA. A double-blind design in a neurosurgery study had the same justification.

Double-blind surgical trials are rare for a couple of reasons. First, there has been a traditional belief that psychological factors play a less significant role with surgical treatments than with drug treatments. Second, they are hard to set up. With double-blind drug trials, the researcher simply has to give the patients a real pill or a look-alike dummy pill. With a surgical procedure, on the other hand, the researcher has somehow to keep the patient ignorant as to whether he or she had the experimental operation or not. With some operations—heart transplants, say—it would be impossible to accomplish this. With other operations, such as those within the abdomen, it is easier to keep the patients in ignorance, especially if the experimental procedure is being compared with a standard procedure that also involves opening the abdomen.

If the procedure is being compared with a no-treatment condition, on the other hand, the surgeon has to be willing to make enough of an incision in the abdomen, in the "control" patients, to mimic the scars that a real operation would cause. And then you have an ethical dilemma, because any kind of incision, even if only skin deep, carries some risk—of infection, especially. The anesthesia that would accompany such a fake operation also carries risks. A patient in the control group is therefore being exposed to a definite (if small) risk without any possibility of a direct benefit to himself or herself. On the other hand, because the effect of the experimental surgery is unknown, the sham-operated patients may actually fare better than those who receive the real operation.

As Stan Fahn and I discussed the matter, we developed a plan that offered the scientific value of a double-blind design and still respected the rights and safety of the individual participants. In our plan, we would tell our volunteers that they had a 50-50 chance of receiving a real fetal-cell transplant or undergoing a sham operation. In either case, they would continue to receive the best established treatment, which is to say treatment with anti-Parkinson's drugs tailored to their individual symptoms. After the operations, each patient would be followed for a length of time sufficient for the transplant (if he or she received one) to exert its hoped-for effects. Then the code would be broken and the patient would be told whether or not he or she had received a real transplant. At that point, any patient who received a sham transplant would have the right to receive a real one.

This arrangement provided a reasonable solution to one of the problems with the notion of a double-blind study. *All* the volunteers would get a real transplant, if they so wished. Thus *all* the volunteers had the possibility of benefiting from a transplant—if the transplant procedure was indeed beneficial. The only uncertainty was whether they got the real transplant right away or later.

Even with this arrangement, we agreed that the risks of the sham operation should be minimal. We decided to do both the real and the sham procedures under local anesthetic, as we had previously. This would avoid the risks associated with general anesthesia. But because the patients would be awake during the procedure, the sham procedure would have to be realistic enough to leave the patient ignorant of whether the operation had been a sham or not. I asked for Bob Breeze's opinion as to how this could be done.

Bob was very clear: the sham surgery must not involve passing any needles into the patient's brain. Needle passes could cause brain hemorrhage, the most serious risk of the operation. The sham surgery could involve making skin incisions and drilling through the skull. The rest would have to be make-believe.

We all agreed with Bob's assessment. We felt that as long as the patient heard and felt real holes being drilled into his or her skull, we could pretend to go through the remainder of the procedure and leave the patient none the wiser. It would be a challenge to our acting abilities, to be sure, but we felt up to it.

As Fahn pointed out to me, a transplant operation is particularly well suited for a "blind" design, because the effects of the operation are not immediately obvious. With other neurosurgical procedures that have been used for the treatment of Parkinson's disease, such as the placement of lesions (localized destruction of a part of the brain) or the implantation of electrical stimulators, the effects are seen right there in the operating room. Thus, with these other procedures, it would be difficult to keep the patient in ignorance of whether the operation was the real thing or a sham. With a transplant, on the other hand, there is no immediate effect during the operation—nothing to clue in the patient as to whether or not cells were really injected into the striatum.

By assigning our participants randomly to a real or sham transplant procedure we could eliminate the role of psychological factors—on the patient's side—in influencing the results of the study. Individual patients would likely have different psychological reactions, of course. Some patients might be convinced that they had a real transplant and therefore feel better; some might be convinced they had a sham transplant and therefore feel no change in their condition, or even a worsening. But there should be no systematic difference between the psychological reactions of the experimental group and the control group, such as might influence the results.

We wanted our study to be *double*-blind, however. Not only the patients, but also the *doctors* who studied them had to be ignorant of whether the individual patients had received a real or a sham operation. How were we to manage that?

Obviously, if Bob Breeze and I performed the transplants, we would know whether each procedure was real or sham. That meant that he and I could not participate in the postoperative follow-up and evaluation of the patients. Stuart Schneck or other neurologists here in Denver could do that job, but it would be a major challenge to prevent leakage of information. How could we discuss a patient with our colleagues on a daily basis—following their ups and downs, adjusting their medications, and so forth—and not inadvertently let slip our knowledge of whether or not the patient had received a transplant? It just didn't seem realistic.

Here was where the New York connection was so opportune.

Stan Fahn and his colleagues formed a tightly knit group that was superbly qualified to follow and evaluate Parkinson's disease patients. They had years of experience in running double-blind drug studies. A large fraction of all the patients they saw were enrolled in such studies, so they were used to being kept in the dark about their patients' treatment status. And New York is 1,500 miles away from Denver; we merely needed to institute some ground rules about communications between the two groups to establish an effective firewall between them.

Here, then, was our plan: Fahn's group would screen volunteers for the study and enroll those who met our criteria. They would perform a baseline evaluation on each patient and then send him or her to Denver, where Bob Breeze and I would perform a transplant operation (real or sham). After the patient recovered from the operation and returned home, he or she would go back to New York from time to time for postoperative evaluations. At the end of the postoperative evaluation period, the code would be broken and the patient would be told privately whether the operation was real or sham. If it was sham, the patient would have the opportunity to request a real transplant. When all the patients in the study had completed their postoperative follow-up periods, the data from all the patients would be gathered together and submitted to a statistical analysis.

The plan was simple in outline, but the details were devilish. Through the early winter of 1992–1993, an ever-accelerating torrent of phone calls, e-mails, FedExed documents, and in-person visits began to flow between Baltimore, Denver, and New York. We wanted to have a grant proposal to present to the NIH the moment that the fetal-tissue ban was lifted, and we expected that to be shortly after Bill Clinton's inauguration in January. Grant proposals have to be very detailed and persuasive. Particularly because we would be asking for millions of dollars—far more than I had ever requested from the NIH. Every question, little or big, that a reviewer might think of, had to be anticipated and answered.

The biggest question was: What were the objectives of the study? As we saw it, the most important objective was to find out whether fetal-cell transplants did in fact alleviate Parkinson's disease. That may sound like a foregone conclusion, given that I've described how several of our patients were markedly improved after receiving transplants. Bear in mind, however, that I am backtracking in time at this point. We wrote the grant application in late 1992, at a time when we had performed transplants on only twelve people, and only seven of these had reached their one-year anniversaries. Some of our most successful transplants, such as those on Bryce TenBroek and Dan Stewart, had not yet been done. Thus, we had much less convincing evidence for the value of the transplant procedure than we did later.

More important, we knew that a string of case histories, even ones with positive outcomes, did not constitute scientific proof of the value of a procedure. For a reminder of this within our own field of Parkinson's disease research, we needed to look no further than the debacle over adrenal transplants. In 1987, Ignazio Madrazo published a glowing account of how patients had been helped by this procedure (see chapter 6), but later it turned out to be worthless. The apparent successes were presumably due to a combination of factors: a possibly real but short-lived benefit of the surgery, plus a large measure of wishful thinking on the part of the patients and their doctors.

The time to do a properly controlled trial is early in the development of a procedure, before large numbers of patients are exposed to the procedure and it develops a life of its own. The history of medicine is full of useless procedures—from bloodletting to bone marrow transplants for breast cancer—that were allowed to become established without proper study and then became hard to dislodge.

Beyond the question of *whether* fetal-cell transplants were beneficial, however, was the question: *Who* did they help? The variability in our results was already obvious after our first two procedures: Don Nelson improved after his transplant, but Anthony Marsh did not. This variability in outcome continued to bedevil our program, as well as similar programs at other centers. Was the variability due to differences in the patients themselves, or was it due to differences in the way the fetal cell survived and developed in each patient?

The most obvious difference between patients was their age. Among the patients we had already operated on by the time we wrote the grant proposal, ages at surgery ranged from 38 to 67. We did not have enough data to say if the younger patients were benefiting more from the transplant procedure than the older ones, so it seemed important to test whether there was indeed a difference in outcome related to age. We therefore decided to design the study with this in mind: we would have equal numbers of patients below and above the age of sixty, and we would look for differences in the results for these two groups.

Among the differences in the way we had performed the transplant procedure, the most notable concerned immunosuppression. In our transplants up to that time, we had given half of our patients immunosuppressive drugs, the other half not. As we were discussing how to organize the double-blind study, we looked carefully at the outcomes for our prior patients and saw no evidence of any benefit from immunosuppression. In fact, the patients who had received the immunosuppressive drugs showed *less* improvement than those who did not receive them.

This difference could have been a matter of chance—we couldn't really conclude that the immunosuppressive drugs *decreased* the effectiveness of the transplant. But we did know that these drugs are potentially harmful to the general health of patients who are treated with them. Therefore, it didn't seem appropriate to expose the participants in the double-blind study (especially the sham-operated patients) to the drugs. We decided not to give immuno-suppressive drugs to any of the participants.

In designing the double-blind study the way we did, we knew that there were certain limitations to the conclusions that could be drawn from it. In particular, we would not be able to distinguish with complete confidence between a beneficial effect of the fetal-cell graft itself and a beneficial effect that might be caused by some other aspect of the transplant procedure.

To make this clearer, imagine how we might have designed the study if we had been doing it on rats rather than humans. In that case, we could have subjected the "sham-operated" animals to a procedure much closer to the real thing. For example, we could have transplanted fetal dopamine cells that had been killed before injection, or that had been taken from fetuses of the wrong age. Or we could have used tissue from the wrong part of the brain. Or, at the very least, we could have passed needles into the striatum and injected nothing. In our actual study we planned to do none of this to our "sham-operated" patients; we would only drill holes in their skulls. Thus, if we saw an apparent benefit in the "experimental" group of patients and no comparable benefit in the "control" group, this benefit could be due either to the survival and function of the transplanted dopamine cells, or (in theory at least) it could be a nonspecific effect of the needle passes or of the trauma caused by the injection.

Working with rats, researchers have done the more elaborate controls just described, and the results have been clear: needle passes alone or injections of cells of the wrong age or from the wrong part of the brain don't confer any benefit. But our study would not provide the same assurance with humans.

One way in which we hoped to clarify the role of the transplanted cells, however, was by means of PET scanning. By this point in our research program we were fairly confident that we could detect and even measure the survival of the transplanted dopamine cells by looking for fluorodopa uptake with a PET scanner. An increase in fluorodopa uptake in the injected portions of the striatum, compared with the uninjected portions or with the uptake levels prior to the transplant, seemed to be a good indication of how well the transplanted dopamine cells had survived the transplant procedure.

If we found that individual differences in the clinical outcome correlated with differences in fluorodopa uptake—in other words, patients did better if more fetal dopamine cells survived—it would be a strong indication that the benefits they experienced were a result of the presence of the transplanted cells, not a nonspecific effect of the trauma caused by the transplantation procedure. Thus, we felt it was important to carefully monitor all our patients by the PET-scanning technique.

To this end, we contacted David Eidelberg. David had been associated with the PET facility at Memorial Sloan-Kettering Hospital in New York back in 1988, when Don Nelson had his preoperative PET scan there. Later, as I mentioned, that unit closed down, and we sent our patients to UCLA or even to the United Kingdom for fluorodopa scans. But by 1992 David was director of a Parkinson's disease unit in a hospital on New York's Long Island—the North Shore Hospital at Manhasset. This hospital had a PET facility and David's group had the specialized skills needed to interpret the fluorodopa PET images of Parkinson's disease patients. David and his colleague, nuclear medicine specialist Vijay Dhawan, joined our team and undertook to carry out all the PET scans for the study. They also assured us that no major changes in the PET scanner were planned for the next few years. Thus we were confident we could obtain exactly comparable scans from an individual patient over the time span of the study, and that any changes in the scans would represent changes in the patient's brain, not in the scanner.

Another difficult question that we had to resolve was: How long should we study patients after their operations before we decoded the results? From a purely scientific perspective, we wanted to make this period as long as possible. After all, the hoped-for benefits of the transplant procedure might take several years to emerge, if the grafted tissue developed only slowly. And, if there were negative consequences of the transplants, these too might take years to show themselves.

On the other hand, there were strong practical reasons for keeping the postoperative study period short. For one thing, grants are usually awarded for a period of four or five years. Given all the time that would be necessary for us to recruit and evaluate patients, and to do all the operations, we might be left with precious little time to follow the patients afterward.

We also felt it would be difficult to recruit patients if the study involved a long postoperative "blind" period. If Stan Fahn told the participants, "You have a fifty-fifty chance of receiving a sham transplant, but if so you have the right to a real transplant six or twelve months from now," we could imagine many participants finding that quite acceptable. But if he told them they must wait several years before finding out whether their operation was real, it might be quite a different matter. People with Parkinson's disease can go significantly downhill in a few years. It just didn't seem realistic to ask patients to remain in uncertainty for such a long period of time.

In the end, we decided on a compromise: we would study each patient for one year after his or her transplant, and then break the code. Thus, we would have double-blind controlled data for only one year after the transplant. But we would continue to study the patients after their codes had been broken. This would be true both for the patients who had real transplants originally, as well as for the sham-operated patients—whether or not they elected to have real transplants at that point. We would then have the opportunity to see how long any benefits of the transplant procedure lasted, and whether there were any late-developing benefits (or harmful consequences) of the procedure.

Another question we had to answer was: How many participants should we have in the study? We knew that we would have four groups: younger patients receiving real transplants, younger patients receiving sham transplants, older patients receiving real transplants, and older patients receiving sham transplants. Given the amount of variability that we expected to see within each group, we decided that we needed eight to ten patients in each group to be able to see significant differences between the groups. We settled on ten patients per group, so that if one or two patients dropped out of a group we could still hope to get meaningful results. That made a total of forty participants.

A final and particularly tricky issue concerned the randomization procedure. Each participant had to be assigned at random to receive a real or a sham transplant—that is the essence of a double-blind trial. But we also needed to make sure that our groups were comparable. In particular, the real-transplant group and the sham-transplant group should have about the same average age, the same proportion of men and women, and the same severity of disease. Otherwise, a difference in the results for the experimental and control groups might be caused by chance differences in age, sex, or disease severity, rather than by the fact that one group got a real transplant and the other a sham.

Now, if we knew at the start of the study who our forty participants were going to be, we could ensure that the groups were comparable with the help of a matching routine. For example, we could take the two youngest participants and flip a coin to see which of them got the real transplant and which the sham; then we could do the same with the two next-youngest participants, and so on. That way we could be sure that the average ages of the people in the experimental and the control groups would end up being similar.

The reality, though, would be very different. We expected that we would be recruiting, evaluating, operating, and following up our patients on an ongoing basis over a period of a couple of years. Some patients would already have been operated on while others were still being recruited. How could we maintain a balance between our experimental and our control groups under those conditions?

To answer this question, we called on the expertise of a Columbia University statistician, Bill DuMouchel. Bill developed an algorithm that allowed us to allocate the patients sequentially over a couple of years and still keep a balance between the groups. The idea was as follows: Let's say that we were at a point in the study where we had already operated on eight patients—four in the experimental group (real transplants) and four in the control group (sham transplants). And let's say the average age of the patients in the experimental group happened to be somewhat higher than that of the patients in the control group. Now let's assume that we had two new volunteers for the study. The algorithm would assign the two randomly to the experimental or to the control group, but it would do so with a bias designed to compensate for the prior age difference: the younger of the two patients would have a somewhat greater likelihood of receiving a real transplant, and the older of the two would have a somewhat greater likelihood of receiving a sham transplant. Thus, as the study progressed, random differences between the groups in age, sex, and disease severity should be evened out.

We accomplished all this—assembling the research team, working out all the details of the proposed study, and writing the actual grant application—between the time of Bill Clinton's election in November 1992 and his inauguration on January 20, 1993. Then we sat down and waited for the hoped-for policy shift.

We didn't have to wait long. Just two days after taking office, on January 22, 1993, President Clinton issued an executive order permitting the NIH to fund research involving transplantation of human fetal tissue into human patients. Within a week of his order, we mailed off our completed application to the NIH, asking for $4.8 million to fund our proposed double-blind study. The application got to Washington in time for the next grant deadline, which happened to be on February 1, just ten days after the executive order was issued. Our preparations had paid off.

Getting the grant application in the mail was not an opportunity to relax. Just a couple of days later, on January 30, Bob Breeze and I did another transplant operation—our thirteenth. Then, on February 3, I went to Washington to testify once more before a congressional committee, this time the Health and Environment Subcommittee of the House Committee on Energy and Commerce.

The subcommittee, chaired by Henry Waxman, was considering legislation that (among other things) would give the NIH specific authority to fund research involving fetal-tissue transplants. This part of the bill was similar to the bill that had been vetoed by President Bush in the previous Congress. President Clinton had just reversed the Reagan-Bush moratorium, of course, but without this legislation he or a subsequent president could reinstate the moratorium at will. It seemed high time to take this matter out of the political arena and put it where it belonged: in the domain of medical research.

Joan Samuelson, Anne Udall, and I gave similar testimony to that we had given to the Senate committee in 1991. I was able to tell the committee about Don Nelson's and Bob Majzler's continued improvement after their transplants. Basically, I tried to convey how important it was for NIH and the federal government to be involved in funding research using fetal tissue.

I was particularly struck by the testimony given by another witness at the hearing, Guy Walden, a Baptist minister from Texas. Guy and Terri Walden had lost two children to a rare genetic disease called Hurler's syndrome. In children with the disease, the lack of one key enzyme leads to dwarfism and mental retardation. Typically, affected children die by their early teens.

Being Fundamentalist Christians, it was the Waldens' practice to seek guidance on all matters from the Bible. On the basis of passages such as "Be fruitful and multiply," the Waldens decided to continue having children in spite of their knowledge that any future children might also be born with the same disease. They did have two healthy children, but their fifth child, Nathan, was diagnosed with Hurler's syndrome while still in the womb.

The Waldens were strongly opposed to abortion, so they planned to accept the challenge of nurturing a third child with the disorder. As it happened, a research group led by Nathan Slotnick (of the University of California at Davis) and Esmail Zanjani (of the University of Nevada at Reno) had been conducting in utero bone marrow transplants in primates, with a view to developing a treatment for diseases like Hurler's syndrome. The Waldens' own doctor put them in contact with Slotnick and Zanjani, who told them that they were ready to try the technique on their first human patient. The Waldens were asked if they would like to volunteer their fetus.

To say that this question provoked some soul-searching on the Waldens' part would be an understatement: they spent many days in prayer and Bible study and in consultation with pastors, lawyers, doctors, and Christian friends before they were ready to give an answer. Eventually they did request the experimental treatment, persuaded in part by the biblical account of the creation of Eve from Adam's rib, which they saw as God's own transplant surgery. In May 1990 Nathan Walden became the first recipient of a fetus-to-fetus transplant in the United States. The transplanted tissue came from a fetus that was aborted to save its mother's life—it was a case of ectopic pregnancy.

Guy and Terri Walden became ardent advocates for overturning the fetal-tissue ban and supporters of research into the medical use of fetal tissue, safeguarded by proper ethical guidelines. It simply seemed wrong to them to allow a child to die because of the abortion issue. "We are pro-life," said Guy Walden at the hearing. "We want to help the living."

Sadly, Nathan Walden later died of causes apparently unrelated to his condition. There was no definitive evidence that the grafted bone marrow cells had survived.

Three months after the congressional hearings I was back in Washington in connection with our grant application. My colleagues and I had been invited to the NIH for what is called a "reverse site visit." In a traditional site visit, an NIH committee visits the institution of a grant applicant to quiz the researchers about their plans, inspect the facilities, and so forth. At the end of the visit, the committee has a closed-door session in which it decides whether to recommend that the grant be funded. In a reverse site visit, the committee stays in Washington and the researchers go there to be quizzed.

With this kind of site visit there is no opportunity for the committee to inspect the research facilities. It is a lot more convenient for the committee, however, particularly if there is more than one grant application to be reviewed. And, as we found out, there *was* another grant application to be reviewed: Gene Redmond's group at Yale had also filed an application by the February 1 deadline, and like us they sought funding for a study of fetal-cell transplants for Parkinson's disease.

When Stan Fahn, Bob Breeze, and I came down to breakfast in our Washington hotel on the morning of the site visit, we found that Redmond's group was already eating. It turned out that they were to see the committee in the morning, followed by us in the afternoon. We were as friendly as the circumstances allowed, but it was a little awkward. After all, the NIH only has so much money in its pot and so these large grant applications are very competitive. I'm sure that each of us thought that the presence of the other team made our own prospects of being funded that much slimmer.

Our group took advantage of the free morning to have an impromptu planning meeting; after all, we didn't have that many opportunities to get together in one place. Toward lunchtime, Redmond and his colleagues returned from the site visit. They looked as if they had been put through the wringer, and I began to wonder how well disposed the committee members were to the whole notion of fetal-cell transplants.

When our turn to meet the committee came, however, everything went smoothly. The committee was chaired by Anne Young, who is professor of neurology at the Massachusetts General Hospital. She and the other committee members asked us a lot of questions, but they were friendly questions by and large, and I began to feel we had a good chance of getting funded. The issue I had felt particularly uneasy about—the double-blind design and the necessity of doing some minor surgery on the control patients—did not seem to raise any red flags. By the end of the day, I felt pretty confident about the outcome of the site visit.

I found out a few weeks later that the committee did indeed recommend that our proposal be approved. In fact, they assigned it the highest possible score, virtually assuring that it would be funded. Redmond's proposal, on the other hand, was not funded. I surmise that the reasons for this were the poor results his groups had obtained to date with their frozen-tissue transplants and the lack of a double-blind design. Redmond later declared himself strongly opposed to the idea of a double-blind trial on ethical grounds. Because of difficulties with funding, Redmond stopped doing fetal-cell transplants. I was sorry to see a principal contributor to this field back away.

Yet another research group, led by Warren Olanow of Mount Sinai Medical Center in New York and Tom Freeman of the University of South Florida at Tampa, submitted a proposal for a fetal-cell transplant study a few months after we did. Their proposal did not include a double-blind design, and it was initially turned down. Olanow and his colleagues eventually revised their proposal to include a double-blind design, and it was funded in 1996.

There were two notable differences between our study and Olanow's. First, they proposed to follow the patients for two years before unblinding the study, unlike the one year we planned for. This lengthened their study—initial results will probably be published in 2002—but it also promised to provide important additional information on the longer-term consequences of the transplants.

The other major difference was that Olanow's group proposed to give immunosuppressant drugs to their transplant recipients. Although our own experience made immunosuppression seem unnecessary, there is still a lot of doubt about this issue, particularly with regard to long-term survival of the transplants. Thus Olanow's study should provide important information that would complement our own.

Even after we heard that our proposal would be approved, it took a long time to get the study started. The official start date was January 1994. However, we spent a lot of time on purely administrative matters, hiring nurses and statisticians, and so forth. Then recruiting participants turned out to be a lot harder than we expected. At the time we had written the proposal, in late 1992, the fetal-cell transplants were much in the news, and we were constantly being approached by potential volunteers. But by 1994 another form of treatment, known as pallidotomy, had seized the headlines. This procedure involves destroying part of a brain region known as the *globus pallidus,* which receives much of the output from the striatum.

Pallidotomy definitely has a place in the treatment of Parkinson's disease—especially for the alleviation of drug-induced dyskinesias. But in the mid-1990s it briefly seemed to be a cure-all and got a lot of media exposure. As a result, interest in the fetal-cell transplants fell off for a while, and some of our patients who had already signed up for the double-blind study dropped out in order to receive a pallidotomy. So we had to recruit patients from outside, by notifying colleagues around the country and by putting announcements in the newsletters of Parkinson's disease organizations.

Yet another reason for delay was the death of Gerald Walster in early 1994. As I described in the previous chapter, he suffered a cerebral hemorrhage during the transplant operation, and this event caused us to changed our surgical techniques: we used thinner needles and aimed them from the forehead rather than from the top of the head. Because of this technical change we had to submit a revised proposal to the NIH.

As is customary with large clinical research grants, the NIH set up a special watchdog committee called a Performance and Safety Monitoring Board, or PSMB, to oversee the study. Ira Shoulson, a neurologist at the University of Rochester with long experience in clinical trials for Parkinson's disease, was appointed chairman. The plan was that there would be a site visit once a year at which we would describe the progress of the study to the members of the PSMB. In addition, we would have to send them a written quarterly progress report. Yet another requirement was that we had to notify the PSMB of any "serious adverse events" that involved participants in the study. A serious adverse event was defined as either death or any medical condition that required hospitalization. The PSMB's role was to monitor these events and make sure they did not reflect any unexpected ill effects of the treatment under study. Serious adverse events did occur, however, and the PSMB intervened in the study, as I'll recount later. But for now the PSMB accepted our revised proposal, and in May 1995 we were finally able to perform our first operation.

CHAPTER 14

Patients

What kind of people would volunteer for our study? To want a fetal-cell transplant at all, at a time when the benefits of the procedure were still uncertain, took an adventurous kind of personality—I had seen that in Don Nelson and the patients who followed him. The unusual conditions of our double-blind trial raised the stakes even higher. Who would agree to the 50-50 possibility of a sham operation, to a year's uncertainty before being told the truth, or to the possibility that any relief from the disease might be postponed for many months or years—if it came at all? I was curious to meet and get to know these forty men and women—strangers to one another at the start of the project, and destined to remain so for most of its length—whose fates would nevertheless be linked more closely than the participants in a reality TV show.

Because of the way the study was designed, however, my interest in the human side of the participants was largely frustrated, at least for the early part of the study. With Don Nelson and the other early volunteers, I had had ample opportunity to get to know them; I saw and examined those volunteers on numerous occasions both before and after their operations. We discussed everything from purely medical matters to all the issues of daily living that made each of their lives unique. But in the double-blind study everything was different. The patients applied to Fahn's group in New York. They were examined in great detail in New York, both before the surgery and after. They had their PET scans on Long Island. They came to Denver only for the surgery.

What's more, because Bob and I knew whether each individual patient received a real or a sham transplant, we had to limit our interactions with the patients to a considerable degree, even on the occasions when we did spend time with them. We couldn't engage in the kind of unstructured conversations that are the mainstay of a good doctor-patient relationship, for fear that slips of the tongue or unconscious clues would give the patients some indication of what they longed to know—whether they had received a real transplant or not. Only after the patients had completed their twelve-month stints in the dark were we able to talk freely with them. And because their return visits were usually to New York rather than to Denver, my opportunities to meet with them were limited.

Eventually I did get to know some of the participants quite well, and Simon LeVay also interviewed several of them for this book. Every one of them is unique, of course, but we selected six individuals to describe in some detail, because their stories are fairly representative of the entire group of forty volunteers.

The first person to be enrolled in the study was "Emily Mason," a teacher and graphic designer living near Toronto, Canada. She contacted Stan Fahn about the study in December 1992, when she was forty-three years old. At that time we had not even submitted our proposal to the NIH, and the formal start of the project was still more than two years in the future. To understand how she was able to grab the first place in line (and ultimately to have the first operation in the study), I need to recount something of her prior history.

Mason, who grew up in western Pennsylvania, seems to have been an adventurous type from childhood. While in high school she enrolled in two foreign-exchange programs, one in Argentina and the other in Sweden. She went to college in Vermont and also attended the Rhode Island School of Design. After teaching in the Vermont public school system for a year, she moved to Montreal to take up a teaching job. In 1983 she and her husband, "Giles," moved to Toronto, where she took up graphical design in the book publishing industry. Outside of her work, Mason engaged in all kinds of adventurous outdoor activities. She enjoyed snowshoeing, ice-skating, and ice-climbing in the winter, and flying and hiking in the summer. She and Giles did several long hiking trips in the Arctic.

Her first neurological symptoms appeared in 1976, when she was only twenty-seven years old. During a visit to a shopping mall she noticed that her left leg began to drag, and she couldn't get it to move properly for several hours. The problem cleared up, but it returned a year and a half later. While snowshoeing in New Hampshire's White Mountains she completely lost control of the same leg and had to crawl on hands and knees for about a mile to get back to the trailhead. Again, the problem corrected itself. The following winter, however, she had a more serious episode. "I was ice-climbing in the Adirondacks," she recalls, "and all of a sudden I'm dead on the rope. I couldn't move, I couldn't get the tools into the ice, and the other climber has to haul me up the cliff. When I get to the top he says, 'I thought you knew how to ice-climb.' I said, 'I do!' " Then a few months later she had another scary experience, while practicing landings in a small plane. She lost control of the plane while approaching the landing strip, and landed very hard. She was unhurt but shaken, and she resolved never to fly solo again.

What first drove her to a doctor, however, was an episode at work in 1980, when she was thirty. She was laying out four pairs of parallel lines for a book design, using a drafting table. It was a simple task that should have taken ten minutes, but it took her all night—she just couldn't get the lines into register. The next day she went to see an endrocrinologist, who immediately suspected Parkinson's disease, because she in fact had the three cardinal signs of the disease: tremor, rigidity, and slowness of movement. Another telling sign that Mason herself had overlooked: her handwriting, once a large round teacher's hand, had shrunk to microscopic proportions over the previous three years. A neurologist admitted her to hospital for tests and, by excluding other possibilities, confirmed the diagnosis.

Parkinson's disease was no alien concept to Emily Mason. Her own father, an obstetrician and gynecologist, was diagnosed with the disease at the age of fifty, when Emily was nineteen. Although he was among the earliest patients in the United States to be treated with L-dopa, he went downhill rather fast and died seven years later. "I had a sense of what was coming," says Emily.

Mason's doctors recommended that she not become pregnant, to spare herself both the stress of pregnancy and the burden of raising children. She heeded their advice. "I did feel a loss that I didn't have kids," she says, "but fortunately I've worked with kids all my life, so that helps. It wasn't a realistic choice to have kids myself, so I made it work. All my time with Parkinson's, whenever I've known that I'm going to have to accept something, I've tried to accept it as if I wanted it, as if I chose it."

Emily Mason's older brother is a cardiologist. Back when he was in medical school he had helped his father get the new miracle drug L-dopa. Now he arranged for Emily to see Stan Fahn in New York. This was in 1980, long before the days of fetal transplants, but Fahn was already a leading authority on Parkinson's disease. At that time, Fahn suspected that the ultimate cause of the death of the dopamine cells in Parkinson's disease might lie with toxic forms of oxygen (oxygen radicals) that are produced by neurons in the substantia nigra as a by-product of the synthesis of dopamine. He therefore started Mason on a course of vitamin E, an antioxidant, with the hope it would slow down the progression of her disease.

Mason didn't notice any benefit from this treatment; in fact, a large clinical study later failed to show any benefit in it. Two years later, her Canadian neurologist put her on regular Parkinson's disease medications. He started her on a trio of second-tier drugs—artane, amantidine, and bromocriptine—and added the heavyweight, L-dopa, two years after that.

Although these drugs did help, her disease progressed. By 1992 her movements had slowed down to the point that every physical activity was a struggle. When she walked, her head lolled downward and to the side, so that she could see nothing but the ground. In the following years the dyskinesias began, and they were soon affecting her whole body. She began to experience unpredictable on-off fluctuations. "Giles and I would go skating," she says, "and I would get maybe five minutes on the ice. Most of the time I was either not moving at all or moving too much. When you're not moving everything's very bleak—it's like the world is going to end. And then all of a sudden you're moving so violently that you're out of control."

Like many people who suffer these wild fluctuations, Mason preferred the dyskinetic state. "When I was 'off' I couldn't even stand up. When I was dyskinetic I couldn't walk in a straight line—I looked like a drunk weaving down the road, with my hands swinging in all directions and my head turning—but at least I could walk, I could get to where I wanted to be in the end."

Mason had given up her job as a graphic designer in 1980, but she was able to continue working as a teacher until 1994. The children, as well as her colleagues, accepted this strangely behaving woman once they got to know her. But going out among the general public was more of a challenge. When people stared at her she would go up to them and say simply, "I have Parkinson's disease."

In 1992 her brother saw the set of articles about fetal transplants, including ours, in the *New England Journal of Medicine*. He called Emily and encouraged her to inquire about it. "You're getting pretty bad," he said. "The window is limited, you're not going to be a good research subject if you get much sicker." So she called Fahn, because she had seen him before and because he had written an editorial that accompanied the *NEJM* articles. In December 1992 she went to see Fahn in New York. She was violently dyskinetic at the start of the appointment, and then she went completely "off." That was about all Fahn needed to know, and he told her about the study that we were planning.

Because we were having long discussions about the double-blind issue around that time, Fahn asked Mason: "How would you feel if you had a fifty-fifty chance of getting a real transplant?" She said "Fine, so long as I could get the real transplant later, if it turned out to be a sham." Thus Mason's input helped convince us that the double-blind design would be acceptable to enough patients to make the study feasible. Fahn told her that she could well be a suitable candidate for the study and encouraged her to stay in touch.

Another way that Mason contributed to the design of the study was via her diary. She had kept detailed records of her daily medications and her on-off fluctuations over the course of each day. She showed Fahn this diary, and besides confirming to him that she would be a good candidate, the format of her diary became the basis for standardized record-keeping of medications and symptoms for all the patients in the study.

Although Mason was now "in the pipeline," she had to wait over two years for her operation, while we struggled with all kinds of organizational matters, hiring staff, and so on. During this period her condition deteriorated. Walking was now often impossible for her, except by peculiar stratagems. "Giles and I had this way of walking where he would get behind me and put his arms around me," she says. "He'd rock his legs back and forth and that would make *my* legs rock. We'd move sideways like some kind of four-legged creature. We called it penguin walking. Or sometimes I couldn't start walking but I could run. I'd be sitting somewhere in a restaurant and I'd run right out of the place. People were quite excited by that—it looked like I was stealing something."

At the beginning of 1994 Mason had to give up work for good. Her condition was now so unpredictable that she could not leave the house by herself. Giles, who is a physicist, started working at home in order to be able to assist her, and he had to take over most household duties such as shopping, cooking, and cleaning. Mason would try to help, but often she made a bigger mess than the one she was trying to clean up. "It's very frustrating going to bed having accomplished nothing that day," she says.

Although Mason is pro-choice on the abortion issue, she wrestled with the morality of using fetal tissue, and was even worried that she might go to Denver for the operation and then not be able to go through with it. Eventually a conversation with a pastor, who pointed out the similarity between regular organ transplants and fetal-tissue transplants, persuaded her that she should go ahead. She came to Denver and had her operation in May of 1995. It was the first of the forty operations in our study.

"Gregory Bennett" was born in Milwaukee, Wisconsin, in 1944. At the age of twenty-one he enlisted in the U.S. Air Force and served for twenty-six years, reaching the rank of lieutenant colonel. He then worked as an insurance company executive in Phoenix. Because of his Parkinson's disease, he took early retirement in 1991, and he and his wife, "Pauline," moved to Connecticut, where they now live. Both Gregory and Pauline have grown children from previous marriages.

Bennett's condition first showed itself in 1989 as a tremor in his right hand. His handwriting, which previously had the regularity that only a Catholic school education can confer, began to shrink. After about three years the tremor spread to his left hand, and soon his right leg was affected, too. Drugs helped, but then he began fluctuating between episodes of tremor and dyskinesia. "When we went out to dinner, he'd be either sitting on his hands or shaking the table," says Pauline. Bennett was never as profoundly disabled as some of my other patients—he continued to drive and travel, for example—but he did require assistance with tasks like tying a tie or buttoning shirts.

Bennett was a patient of Paul Greene, the colleague of Stan Fahn who was responsible for most of the patient evaluations in the double-blind study. Sometime in late 1994, Greene suggested to Bennett that he consider enrolling in the study. Bennett was not very interested: the idea of anyone doing something to his brain did not appeal to him. But Pauline was more enthusiastic. "I was determined that he was going to do it," she says. "I didn't see any other option." Eventually Gregory came around to the idea.

The use of fetal cells was not an issue for either of them:

although both Gregory and Pauline had strong Catholic upbringings, neither of them are particularly religious, and they are certainly pro-choice on the abortion issue. They were more divided on the question of the double-blind design. Gregory was comfortable with it, but Pauline didn't like the idea. "I really felt that it was quite extreme to do that for something as significant as brain surgery," she says. "But it was their bat and their ball: if you wanted to take advantage of the opportunity you had to sign on the dotted line, or you weren't in the study." So in the end she supported Gregory's decision to go ahead, and he had his surgery in September 1996.

Sid Howard experienced the first symptoms of Parkinson's disease in 1988, when he was sixty-eight—more than forty years older than Emily Mason when she had her first symptoms. Howard was an economist—he had been on the NATO international staff in Paris during the Kennedy and Johnson years and had advised member nations on the U.S. plan for the buildup of the NATO alliance. Outside of his work, his main hobby was tennis, which he played passionately and skillfully.

While growing up on Manhattan's Lower East Side, Sid met his future wife, Nee; they married in their early twenties, and they now have three children and three grandchildren. After Sid's retirement they moved to Southern California. Like Robert and Chris Orth, they live within a few paces of the Pacific Ocean, but in more upscale surroundings: they recently designed and built their own home on a beachside bluff and filled it with Chinese paintings and sculptures they have collected over the years.

It was in China, in fact, that Sid's condition was first diagnosed. Sid and Nee were walking one day with a small group of traveling companions. One, a doctor, motioned Nee to drop back, and said to her, "I think your husband has Parkinson's disease." He had come to this conclusion because Sid dragged his feet slightly while walking and had a certain hesitancy to his speech. He also mentioned what he called Sid's "sunburst smile." By this phrase, he may have meant that Sid's smile came as a surprise, because his face was usually rather immobile.

A few months later, neurologists in the States confirmed the diagnosis. Sid was started on Sinemet and did well: he was able to continue playing tennis, for example. But by the mid-1990s he was developing troublesome dyskinesias that affected his mouth, tongue, jaw, and neck. His speech became slow and difficult to understand, his handwriting deteriorated, and his mobility became more and more impaired. In 1996 he had to finally give up playing tennis, which he experienced as a major loss.

Howard frequently asked his neurologist about the possibility of novel therapies. In 1994, hearing about our study, the neurologist recommended that Howard visit Stan Fahn, which he did. Howard seemed an excellent candidate. At seventy-four, he was older than most other applicants, but we had set the upper age cutoff at seventy-five, so he just made it under the wire. In spite of his age, he did not have major health problems besides his Parkinson's disease, which might have been reason to exclude him from the study.

Neither Sid nor Nee had moral reservations about abortion, so the use of fetal tissue posed no problems for them. They were also comfortable with the double-blind design of the study. Nee was already a participant in another NIH study, so the rigmarole of these trials was familiar to them. In fact, they were able to synchronize their checkups so that they could travel to the East Coast together. Recently, Sid Howard told us that he would have participated in the study even if there had been no promise of a real transplant for the patients in the sham-operated group. "I guess I've got altruistic motives," he said. "People don't volunteer for these things just to get a benefit for themselves." He had his surgery in February of 1996.

Like Sid Howard, Jacob Celnik grew up on New York's Lower East Side, but he wasn't born there. Celnik is Jewish, and at the age of three he escaped from Nazi Germany with his parents, just as World

War II was about to break out. His grandparents perished in the Holocaust.

At school, Jacob (who usually goes by Jack) proved himself an assiduous and gifted student. He had a particular interest in Jewish culture and history, and by the age of fourteen he had already mastered Judaic law at a college level. But he also had a bent for science and ended up entering graduate school in physics at Columbia. The death of his mother, and the resulting turmoil in his family, prevented him from formally completing his Ph.D., but in spite of that he had a productive career as a nuclear physicist. He is the author of fifty scientific papers, many having to do with radiation shielding. He worked extensively for NASA on techniques for protecting astronauts from radiation in space, and he also invented a robotic rover for work in high-radiation environments. Another interest was statistics; he helped develop what is now a popular computer-aided statistical technique called the Monte Carlo procedure.

Celnik married at the age of twenty-four and the couple had five children. His wife died when the youngest child was only five. In that same year, when Celnik was forty-nine; he was diagnosed with Parkinson's disease. Thus he had to cope with his early bereavement, the demands of his work, the responsibilities of a single parent, and the burden of a progressive illness. "It was difficult sometimes," he says—which is doubtless an understatement.

Celnik's illness started as a weakness in the left hand, and it gradually extended to his left arm and left leg. A tremor soon appeared. As the years went by, his gait became hesitant, then slow and shuffling. His writing deteriorated, his speech became slurred, and he had to stop driving. Fiddly manual tasks, such as buttoning his clothes, became difficult, and sometimes his children had to help him with eating.

L-dopa was the main medication Celnik took to combat the illness, and it had always relieved his condition considerably. In the course of his illness he has been to see quite a number of neurologists, many of whom have started him on additional drugs. "I've taken everything under the sun," he says. Many of them helped, but as the years went by he developed dyskinesias that interfered with his work, especially his teaching. "It's difficult to stand up in front of a class and have your hands go crazy," he says.

One of the neurologists Celnik visited was Dr. Melvin Yahr, of Mt. Sinai Medical Center in New York, who has done considerable work on Parkinson's disease. Yahr mentioned the fetal-cell transplant study being conducted by his Mt. Sinai colleague, Warren Olanow. Celnik liked what he learned about the fetal-cell transplant procedure, but after he looked into both Olanow and our studies, one factor persuaded him to go with ours—the fact that we were not using immunosuppression. So in early 1996 he applied to enroll in our study. The initial evaluation showed that he met the criteria; he was accepted, and he had his operation in January of 1997.

Concerning the use of tissue from aborted fetuses, Celnik says that his own extensive knowledge of Jewish law and traditional moral teachings led him to believe that it was permissible to accept such a transplant. Several other authorities he consulted concurred with his opinion.

James Ross, a retired mathematics professor, was born in 1939 and grew up in Minnesota, attending college and graduate school at the University of Minnesota. In 1969 he obtained a tenure-track position at San Diego State University, and became a full professor there in 1979. He retired because of his Parkinson's disease in 1999. He and his wife, Gail, still live near San Diego. As befits a mathematician, his favorite recreation is Go, the intensely cerebral board game.

As with Gregory Bennett, Ross's illness began with the classical Parkinsonian tremor. He was forty-two years old at the time, and the tremor first affected his left leg. He visited his HMO, where he was examined by a nurse who recommended that he take a course for people with psychological problems. But Ross himself suspected he had Parkinson's disease, and eventually he found his way to a neurologist who confirmed that self-diagnosis.

Ross's disease progressed in a fairly typical way, gradually interfering with his handwriting and with physical activities such as tennis, his favorite sport. His gait was particularly impaired: he walked on the balls of his feet, and typically he would take a few short steps and then break into a run, as if trying to save himself from falling forward. Sometimes he could only end the run by crashing into a wall or doorjamb, or by falling. He began to experience the typical Parkinsonian freezing episodes, especially when passing through doorways when there were other people in the vicinity. By the early 1990s he was also suffering from dyskinesias that affected much of his body.

Around 1993 Ross became interested in the possibility of receiving a fetal-cell transplant. After doing some research into the various options, he wrote to me, and later came to visit me in Denver. Like Jack Celnik, Ross was particularly motivated by the fact we were no longer using immunosuppressive drugs, whereas the other groups were still doing so. Ross felt strongly that he did not want to take these drugs. Part of his reason for this may have been that he had a history of hypertension, which can be aggravated by immunosuppressive drugs such as cyclosporine.

When Ross initially saw me, he was interested in having the fetal-cell transplant as a regular private patient. When I explained the likely costs of the procedure, he decided to volunteer for the double-blind study. He went through the initial evaluation procedure here in Denver, but then repeated the process in New York like the other study patients. The results of this evaluation showed that he was not severely enough affected by his disease to meet the criteria for the double-blind study. Indeed, in spite of his gait problems and his dyskinesias, he was still able to hold down his job, since his intellect was unimpaired. We therefore told him that we could not accept him at that time, but invited him to stay in touch.

"I had to deteriorate a little more," as Ross put it recently. "I did deteriorate, and then I was bad enough." Two years after we first examined him, Ross's test results were well within the range for acceptance into the study, and he was enrolled. He had his operation in August of 1997.

James and Gail Ross had differing views on the ethics of using fetal tissue for transplantation. Gail is a Christian who believes that abortion is wrong, and she says that she would not herself have a fetal-cell transplant under any circumstances. James was not a Christian at the time he was considering a transplant. "Most of the people I talked to about it were academics," he says, "and they were in favor of going ahead." Gail respected his decision to enter the study. Since the time of his surgery, James has also become a Christian, but that has not affected his views on the matter. "It's not a straight religious issue," he says. "There are plenty of religious people who think that these operations are acceptable."

Jacqueline Winterkorn was born in Queens, New York, in 1947. Her life has been one of high academic and professional achievement: she has an A.B. from Columbia, a Ph.D. and an M.D. from Cornell, and is now professor of neurology and neuroscience, and also of ophthalmology, at Cornell University Medical College at New York Hospital. She has two grown children and lives with her third husband, David, in rural Connecticut.

Winterkorn's research and clinical expertise deals with disorders of the visual system, including the brain pathways that analyze visual signals and control the movements of the eyes. Before coming to New York Hospital, she had a position at the North Shore University Hospital on Long Island, where the patients in our study received their PET scans. She even collaborated with David Eidelberg, the neurologist there who is now a member of our research team. But this connection to our study was coincidental and had nothing to do with how she ended up participating in it.

In 1988, at the age of forty-one, Winterkorn noticed an occasional "pill-rolling" tremor in her left hand. With her training, the possible significance of this could not escape her. "I looked at my hand," she told us recently, "and I said to myself 'My God, I've got Parkinson's—this can't possibly be!' " In addition, a colleague at work asked her whether she had hurt her left leg. The question drew Winterkorn's attention to something she hadn't noticed—her left ankle would go into a repetitive spasmodic motion known as clonus. Winterkorn went to see a neurologist, who noticed an additional symptom: she was not swinging her left arm in the usual way as she walked. But he disagreed with Winterkorn's self-diagnosis. "It's not Parkinson's disease," he said. "You're too young."

A year later Winterkorn went to see Stan Fahn at Columbia, who examined her and said, "Of course, it's Parkinson's." As with Emily Mason, Fahn started Winterkorn on vitamin E, along with another drug, selegiline (also called Deprenyl or Eldepryl), which interferes with the breakdown of dopamine, thus increasing the levels of dopamine in the brain. Unlike vitamin E, which has since turned out to be ineffective, selegiline can improve the symptoms of Parkinson's disease, although the effect is not a dramatic one.

Winterkorn, however, did not notice any benefit from these medications; her condition worsened steadily. By the early 1990s, she was having increasing difficulty in walking. "I never knew, when I stood up, whether I was going to walk or whether I would freeze and fall over," she says. If she did walk, it would be a stiff shuffle, or a series of small, hurried steps. She and her second husband, Thomas Meikle, liked to take five-mile walks on Saturday mornings. "I remember that I'd get very annoyed if Tom would stop to look in store windows," she says, "because I was never sure when we started off again if I'd be able to get the rhythm back. Sometimes I'd be hobbling so badly that we'd stop for a drink, and regardless of whether it was alcohol or a Coke, when we stood up to walk away I'd be walking smoothly again." Thus Winterkorn was suffering from on-off fluctuations even though she had not yet taken any L-dopa, the drug that is sometimes blamed for provoking these fluctuations.

Around 1992 Winterkorn did start on L-dopa, which helped immediately, increasing the amount of time she was "on" and reducing her stiffness. But after a couple of years she developed dyskinesias. These were so pronounced that she came to have mixed feelings about having taken the drug—she wished she had tried one of the dopamine agonists first. As well as affecting her limbs, the dyskinesias also involved her tongue and throat. Her speech became difficult or impossible to understand over the telephone. This caused major problems in her work, for she often needed to report by telephone about patients who had been referred to her.

Another symptom that impeded her work was a deterioration in her handwriting. She lost the ability to take notes while seeing patients and had to rely on her memory instead. She even had to get a stamp made that she could use in place of her signature.

Besides the problems at her work, Winterkorn faced many other difficulties dealing with life in New York. "The subway is not a secure place for someone with Parkinson's disease," she says. "Taxi drivers assume you're drunk and refuse to pick you up. You go into a store and the guards start to follow you around. And when you reach the check-out line, your hand is not working properly to reach into your purse and pick out coins, so the lady behind the counter talks over your head to someone behind you, and they raise their eyebrows at each other." She sums it up with an uncharitable aphorism: "If you're in a place that's a big rat race, you tend to meet rats." She and her husband spent as much time as possible at their second home in northwestern Connecticut, where they developed a large nature garden.

From time to time, Winterkorn asked Stan Fahn whether there were other treatment options open to her. For several years he was discouraging, but in 1996 he called her and told her about our study, which at this point had been under way for a couple of years. "We're starting to get results from the first patients," he said. "It looks like they're doing well, and there haven't been any complications. This is your best hope, there isn't anything else on the horizon."

Winterkorn was dubious at first. "To someone who had been a neuroscience graduate student in the 1970s," she says, "the idea that you could put fetal cells into the brain and get a good result seemed preposterous. But when I read some of the recent papers, it looked a lot better—it seemed like my best hope. The alternative was a pallidotomy, and it didn't strike me as a good idea to make a lesion [area of damage] in my brain when I already *had* a lesion in my brain. If the fetal cells worked this could be more like a repair." So she signed up for the study, becoming the fortieth and last participant.

Winterkorn was not a particularly religious person at the time, and although she has become more observant over the last couple of years, that has not changed her views about the morality of fetal-tissue transplants. "The notion of taking an aborted fetus and making it useful to people who are already living—people who are already having relationships and trying to have a good effect on the people around them—is a very moral thing to do."

Around the time Winterkorn signed up for the study, she was struck with another blow: her husband was diagnosed with colon cancer. He died about a year later, in July 1997. By this point Winterkorn had completed her neurological and psychological tests for the study, and she had her preoperative PET scan a week after her husband died.

It's hard to imagine how difficult life must have been for her at this time. "I was mourning my husband's death, I had to leave the townhouse we lived in, because it belonged to Tom's employer [he had been president of the Josiah Macy Foundation], I was commuting to work from Connecticut, where I supposedly was going to live with my retired husband, but now as a widow. I didn't drive so I was taking public transportation—a two-hour trip to get to Long Island where my practice was, then teaching with great difficulty at New York Hospital. And I had to wait six months for my transplant. They had other people ahead of me in line, and they only did one surgery a week, and sometimes they were traveling and didn't do surgery, so it dragged on and on, and I was on tenterhooks the whole time." Finally, in January 1998, Winterkorn did come to Denver and get her surgery. The operation on Winterkorn, our fortieth patient, rounded out the first stage of our study.

At some point before they had their surgeries, all the participants in the study underwent a preoperative evaluation. For this purpose, each patient came to Stanley Fahn's clinical research unit at Columbia-Presbyterian Medical Center on the Upper West Side of

Manhattan and stayed there for several days as an inpatient. During this time, he or she was subjected to about as thorough a neurological and psychological evaluation as it is possible to imagine. "The five-day torture" was how Gregory Bennett's wife, Pauline, described it. Most of the patients themselves were pretty good-humored about the whole thing. It was an imposition, certainly, but when you're investing five million dollars in a study you want to get your money's worth. We felt we had to know these patients as no Parkinson's disease patients had been known before. Only then could we make the best sense of whatever changes we saw after the transplant surgery.

The evaluation was carried out primarily by two registered nurses, Sandra Dillon and Hal Winfield, who are highly skilled in neurological testing. Most of the testing was done within the framework of the unified Parkinson's Disease Rating Scale or UPDRS, which was developed some years ago under Stanley Fahn's leadership. One component of the UPDRS testing is the "motor component," which looks at the patient's movements. This includes a set of tests such as having the patient hold one hand out and rapidly alternate between the palm-up and the palm-down position. Each arm and leg is tested separately. The patient is also asked to perform tasks like getting out of a chair. Speech, posture, gait, tremor, and rigidity are also tested. Every test is scored from zero, meaning that the patient's performance is normal, to four, which means that the patient cannot perform the test at all. The scores from all 27 tests are combined, so the final score ranges from zero (completely normal) to 108 (maximally impaired). The motor component is very important because, after all, Parkinson's disease is a disorder of movement, and the motor component provides an objective measurement of how badly the patient's movements are impaired.

We also wanted to find out how the patients were doing in all the routine activities of their daily lives—activities that can be severely impaired by Parkinson's disease. We used two sets of questions on this topic. One was the "Activities of Daily Living" component of the UPDRS. In this component the patient is asked about specific activities such as cutting food, dressing, and turning in bed, as well as problems that can affect Parkinson's disease patients such as falling and freezing. There are thirteen items in this component, so the total score ranges from zero (normal) to 52 (maximally impaired). We also used a second set of questions about the activities of daily living called the Schwab and England question set, which rates the patient's overall state on a percentage score, where 100 percent is completely normal and zero percent means bedridden and incontinent. With these tests, we asked the patients both about their performance in the "on" state and in the "off" state (for example, on awakening in the morning), so we got separate "on" and "off" scores, just as with the motor UPDRS.

Finally, we wanted to assess the patients' cognitive and emotional state. The UPDRS has a component called Mentation, Behavior, and Mood, which assesses the patient's motivation, memory, and orientation in time and space, as well as the occurrence of thought disorders and depression. The total score here runs from zero (normal) to 16 (maximally impaired). In addition, Dillon and Winfield gave the patients another well-known test called the Mini Mental Status examination. This assesses the patient's memory and cognitive state with questions like "Count backwards from 100 by sevens," and so on.

One of the major problems in assessing treatments for Parkinson's disease is that a patient's condition can fluctuate tremendously from day to day or from hour to hour, and is strongly influenced by the drugs the patient is taking, especially L-dopa. Thus, in order to get a reliable baseline representing a patient's condition prior to the transplant surgery, we had Dillon and Winfield test each patient on at least six occasions in the "off" state (after the patient had gone without L-dopa and other medications for twelve hours), and on at least six occasions in what we call the "best-on" state: this was anytime during the day when the patient could move the best.

Of course, the patients' performance on the motor component of the UPDRS was better in the "on" than in the "off" state. In fact, one of the criteria for acceptance into the study was that a patient's UPDRS score improved by at least 33 percent in the "on" state compared with the "off" state. This criterion ensured that all our patients were responsive to L-dopa, and therefore that they probably had classical Parkinson's disease and not some other movement disorder. Another criterion for acceptance into the study was that the patients scored within the normal range on the Mini Mental Status exam. It's been our experience that patients with cognitive impairment do not do well with fetal-cell transplant surgery; their mental state can become worse, at least temporarily.

Another test the patients had to endure during their preoperative evaluation was a PET scan. This was done by David Eidelberg's group at the North Shore Hospital on Long Island. I'll have more to say about that test in chapter 17.

The preoperative evaluation at Fahn's unit was to be repeated on a regular basis after the operations: every four months during the first postoperative year and every six months thereafter. These postoperative evaluations had one important extra component, which we called the Global Rating Scale. This was simply the patient's own assessment of his or her condition compared with before the surgery. He or she was asked to check one of seven boxes: Much worse than before the surgery, Moderately worse, Slightly worse, Unchanged, Slightly improved, Moderately improved, or Greatly improved. These answers were given numerical values from minus 3 to plus 3.

It was Fahn's idea to include this self-evaluation. In spite of my reservations, the Global Rating Scale was made the "primary outcome variable" of the study, meaning that it was to be the principal measure of whether or not the transplants were effective. I would have preferred to have the UPDRS as the primary outcome variable, because it is an objective and widely used assessment of the severity of a person's Parkinson's disease. The Global Rating Scale, on the other hand, was a subjective measure that could be influenced by all kinds of factors, and it was made up for this particular study rather than being a standard test.

Fahn favored the scale because he felt that a radically new treatment for Parkinson's disease, such as fetal-cell transplant surgery, should only be introduced into practice if its superiority to other treatments was obvious to all concerned. I, on the other hand, felt that fetal-cell transplant surgery was still under development, so that even more modest benefits, if they could be objectively verified, were of great importance. Much later, I came to regret not having fought harder to make the UPDRS "off" score the primary outcome variable.

CHAPTER 15

The Real and the Sham

By May of 1995, when we did our first operation in the federally funded study, Bob Breeze and I had done more than twenty fetal-cell transplant surgeries. Over the seven years since Don Nelson's first operation, we had gradually improved the speed, safety, and efficacy of the procedure. It was becoming routine. But now we were being asked to add a whole new dimension—secrecy. Of the next forty patients, one-half were selected, by the toss of a digital coin, to get a real transplant; the other half were to get sham surgery. And no patient could be allowed to find out the truth—not for twelve months, at least.

Here's how we did it. On the morning of each operation Bob met the patient in his or her room, and mounted the metal "halo" on the patient's head. As always, local anesthetics dulled the pain, but it was certainly the most uncomfortable part of the procedure for everyone. At least it was over right at the beginning of the day. After that, the patients were free to walk about, albeit with this strange heavy object fixed to their skulls and restricting their vision. The patients' families tried to ease the awkwardness with humor. Emily Mason's husband

[180]

Giles referred to the apparatus as her "crown jewels"; Jack Celnik's sons called him "The Man in the Iron Mask."

Having fixed the halo, Bob went with the patient to the basement for an MRI scan. Once he got the scan back, Bob went through the calculations to establish where the holes needed to be drilled and the angles and distances for the needle tracks. Because the holes would be drilled very near the frontal sinuses—air-filled spaces in the skull that communicate with the back of the nose—Bob needed to make sure that the drill would miss the sinuses themselves. Otherwise, there was some risk that bacteria that normally inhabit the sinuses would spread to the brain.

Bob did all this without knowing whether the operation was to be a real one or a sham. At that point, only two people in the world knew. One was Bill DuMouchel, the statistician back in New York whose computer was, in a sense, running the whole show. The other was myself. A few weeks earlier, Bill had sent me the names of two or three patients—the next ones in the pipeline—with instructions as to whether each patient was to get a transplant or a sham operation. I then scheduled the operations for the individual patients, based on the availability of fetal tissue. If we had tissue available, I might schedule a patient the computer had selected for a real transplant. If tissue was not available, I might schedule a patient who was due for a sham transplant. But I kept the order random-looking; it never alternated regularly between real and sham surgeries, nor were there any long sequences of one kind of surgery or the other.

We had agreed to carry out all the surgeries at the same time and day of the week—we settled on Thursday mornings. Bob Breeze would leave those mornings clear unless I told him that we had no transplant scheduled. His job was to oversee the workup of the patient and the preparation for the operation. Since he did not know whether the surgery would be a real or a sham procedure, there was no way he could communicate to the patient—by how carefully he fixed the halo, for example—what was in store.

When the patient was ready in the operating room, I would come down from my lab, carrying a cooler. In the cooler was a rack with four test tubes containing a buffered salt solution and tiny fragments of fetal tissue. If a real transplant was scheduled, that tissue consisted of fetal substantia nigra—the 2- by 3-millimeter chunks that I had dissected up to four weeks earlier and kept bathed in a nutrient solution. If a sham transplant was scheduled, there was still tissue in the box, but it was waste tissue of some kind. After I put the cooler down near the operating table, Bob and I left the room to wash our hands.

While we were scrubbing up and well out of hearing of the patient, I would say to Bob: "So what kind of operation are we going to do today?" The only way he could give me the correct answer, aside from pure chance, was if he could glean some clue from my behavior or mannerisms. Remarkably, he did guess correctly for the first several surgeries in the study. I think he got nearly all of the first ten surgeries correct, although we didn't keep records.

How did he do it? When I asked him, he'd say things like "You're a little more nervous today, so I think we're going to do an implant," or "You're right on time, which means it's a sham. If it's an implant you're usually a few minutes late." When I found out what was clueing him in, I changed the relevant behaviors. In addition, the whole procedure became more and more routine, so there was less reason for nervousness. By about the tenth surgery Bob's success rate in guessing the kind of surgery went down to chance levels.

Having heard Bob's guess, I would tell him what was actually scheduled, and we would go back into the operating room to get the patient ready for surgery. Unlike Don Nelson, who watched the whole procedure via reflections in a cabinet window, these patients would not be allowed to "peek," so we covered their view with surgical drapes that we laid across the halo. Now we only had to worry about what sounds and feelings the patient might experience, plus any clue that he or she might gain from the overall duration of the procedure.

Besides Bob, the patient, and myself, there were three other people present in the operating room during the operations. Two were scrub nurses, and the other was the anesthesiologist. (In the early operations, such as that on Nelson, we had had a surgical resident and several other people in the room, but we had gradually dispensed with them over the years.) The scrub nurses were members of our team, and they understood the importance of not breaking the "blind."

With the anesthesiologist it was a little different. Anesthesiologists are provided by the hospital: you never can be sure who you are going to get on a given day. We felt that there was a significant risk that an anesthesiologist might unintentionally break the blind, either during or after the operation. We therefore arranged things so that the anesthesiologist was kept in the dark, just like the patient. This we did not just by limiting our communications, but also by hanging surgical drapes across the patient's head. The drapes effectively divided the operating room into two zones. One zone included the top of the patient's head, and this was where Bob and I were operating; the other included the rest of the patient's body, and this was the anesthesiologist's domain. The strategy worked well—to my knowledge, no anesthesiologist ever found out what kind of operation we performed on a given patient.

The operating room that we used has glass windows in the entry doors, through which personnel not involved in the study might get a peek at what was going on. To prevent this from happening, we covered the entire windows with cardboard. We also put a notice on the door prohibiting visitors. In spite of this, we heard the door opening at one point during the first operation; a man in dark glasses and carrying a cane tapped his way in. It took me a moment to recognize him as the senior member of the anesthesiology faculty. He said, "I'm your double-blind anesthesiologist." We had a little laugh at that. After that we had no more unwanted visitors.

Bob began the operation by mounting the steel arc onto the halo and using it to determine where the skin incisions should be made. Off came the arc again, and Breeze anesthetized the skin around the incision points. Then he made the incisions, trying to keep them in one of the patient's "frown lines" so that the scars would be less obvious later. Having made the incisions and exposed the outer surface of the skull, it was time for him to take his hand drill and make the four holes, two on each side of the forehead and each one-eighth of an inch in diameter. The drill was held in position by the steel arc to make sure that the hole would be in precisely the right place.

While Bob was doing this, I was getting the tissue ready. Whether it was real substantia nigra or waste tissue, I did exactly the same thing: I loaded the chunk of tissue into a glass pipette, and then slowly extruded it from the narrow end of the pipette by pressing on the plunger of a syringe that was attached to the other end. The air pressure forced the tissue slowly out of the narrow end in the form of a "noodle": this could be several centimeters long, depending on the exact size of the specimen. The noodle was now floating in a petri dish full of nutrient solution, and it was sitting on an ice block to keep its metabolism in a suspended state.

Now for the first time the scenarios of the real and the sham operations diverged. If it was a real operation, Bob took a sharp-sided needle and cut a slit in the dura mater, the brain's protective membrane. If it was to be a sham operation, on the other hand, he merely went through the motions of cutting the dura. Often he would actually scratch the dura with the little needle, but he would never cut it.

Some patients said that they felt the dura being cut. However, there didn't seem to be any reliable connection between whether they felt anything and whether the dura was actually cut. For example, both Emily Mason and Jacqueline Winterkorn said they felt some pain, but one of them had a real transplant and the other a sham. Because of her training, Winterkorn knows as much about the brain as I do, and at that point in the operation she said, "I feel that sort of sickening C-fiber pain—are you going through my dura?" (C-fibers are extremely fine nerve fibers that are responsible for a dull, poorly localized pain or ache; larger-diameter fibers carry signals related to sharp pain such as that caused by a pinprick.) Bob said, "We purposefully tickle the dura, so everyone feels that." Still, I made a mental note to exclude neurology professors from future double-blind studies!

Bob next took the triple-gauge special transplant needle that Trent Wells had made for us and guided it into the brain with its rounded stylet in place, using the halo and its carefully positioned guide holes. At this point, the calculations on the MRI scan determined where the end of that needle would go. Bob had calculated it to be at the far end of the putamen. We always started with the left lower needle pass, since that is the area of the putamen most likely to control speech. It is the riskiest of the four needle passes if a hemorrhage should occur, so we wanted to make sure it went well before moving on to the other three sites.

I now took the long, very thin-walled needle that would actually carry the tissue and attached a microliter syringe to the needle hub. This kind of syringe can accurately dispense the tiny amounts of material that make up a neural transplant. I filled the needle and syringe with the buffered salt solution, and then by pulling on the plunger of the syringe I gradually sucked the strand of tissue up into the barrel of the needle. The total volume of the tissue and fluid was about 20 microliters. To put this into perspective, one drop of water from a standard eyedropper is about 50 microliters, about two and a half times more than one tissue transplant. Once the entire strand was inside the needle, Bob and I now worked together to precisely position the tissue strand in the brain.

Bob withdrew the long rounded stylet that had protected the patient's brain from the sharp edges of the guide cannula during its insertion. I then slipped the long needle with the tissue down the guide cannula. Now the careful placement of the tissue could be done. The needle was sitting precisely in the patient's putamen.

Bob and I coordinated the tissue deposit. He began to withdraw the needle assembly as I began to inject tissue into the brain. The typical patient has a putamen that is about 40 millimeters long, which is about an inch and a half. We wanted to spread the tissue over that length. Since my syringe held about 20 microliters, I would press the plunger to release one microliter as he moved back 2 millimeters. To coordinate this effort, Bob would call out the measurement every millimeter. "Forty, thirty-nine, thirty-eight . . ." would become a rhythm that the patient and the staff got used to. Once he got to zero, we had completed the first of the four transplants. A strand of tissue had been laid down precisely along the

40-millimeter length of the putamen. We paused for two minutes at that point to allow any pressure that might have built up to dissipate. Bob then withdrew the needle assembly from the brain as I injected a small amount of buffer to fill the void that was left as the needle came out. Bob sutured the incision closed as I took the needle over to the petri dish to make sure all the tissue was delivered and that no bleeding had occurred. I then asked the patients to talk to us and to move their arms and legs. Sometimes they had dozed off, so we had to wake them to make sure that everything was okay. Bob and I now repeated this sequence for three more injections. Almost invariably, each transplant site took fifteen minutes to do, so we could say with some confidence that the transplant part of the operation would be over in an hour.

If this was a sham operation, nearly everything was the same. Bob drilled holes in the skull, touched the dura with a needle, and positioned the guide cannula into the carrier on the halo. He moved the tip of the cannula close to the hole in the skull, but did not touch it. I extruded tissue, but did not load the needle. The whole system just contained air. With the guide cannula not in the brain and the tissue not in the needle, we moved ahead with our standard method. Bob pulled the inner stylet, and I placed the transplant needle into the guide cannula. Bob called out the numbers as I carefully pressed on the plunger of the syringe, staring intently at the markings of the air-filled syringe. From the beginning, both the real transplant and the sham surgeries were carried out with identical rhythms. We were dedicated to having the sham operations performed as perfectly as the tissue transplants.

When we had finished, Bob placed Band-Aids over the sutured skin incisions. Then he took off the halo and we wheeled the patient to the recovery room. The whole procedure took the same amount of time, whether it was a real transplant or a sham.

Now we had the real acting job, which was to discuss the procedure with the patients in the recovery room and their families in the waiting room. The patients were invariably happy to have the operation finished but were hungry and wanted their L-dopa. After getting the patients set with their medications and appropriate food, we went to see the family. Bob was always able to give good news: "The operation went exactly as planned, and Emily is doing fine in the recovery room." Reassuring but incomplete, and accepted good-naturedly by the anxious relatives.

Given that we would be making a total of eighty needle passes in the course of the study—four passes in each of the twenty real-transplant patients—we knew that there was a significant chance of something going wrong. The most likely mishap was a hemorrhage. We had carefully planned for what we would do in such a situation, and this was a good thing because we did in fact have a hemorrhage in one of the operations. It occurred during the last of the four needle passes in that patient. We recognized the hemorrhage as a streak of blood left over in the needle as I checked it after the transplant. All other transplants in all our patients had been performed without a hint of blood. It is remarkable that neural transplant surgery, at least in Bob's hands, could be bloodless surgery.

Following our plan, we waited a few minutes to see if the patient had any signs of weakness in any limb or slurred speech. If any signs of a stroke developed, our plan called for us to break the blind, tell the patient what had happened, and concentrate on treating the stroke. In fact, however, there were no ill-effects: the patient had no change in neurologic condition and had no complaints. The MRI after the operation did show a small amount of blood in the frontal lobe, away from the region we had transplanted but along the needle track. Because asymptomatic hemorrhage detected only on MRI scan was specifically excluded as a reason for breaking the blind, we did not do so. Of course, we immediately informed the NIH oversight committee about this complication, and they agreed it did not requiring breaking the blind. The patient finished the twelve months of blinded follow-up uneventfully. When finally informed about the mishap a year later, the patient said, "I had no idea that anything like that had happened."

It turned out that about 75 percent of the patients came out of the surgery believing that they had had a real transplant, and the remainder were uncertain but hopeful that they had had one. Few if any patients thought they had had a sham operation. So why was there such a strong bias in favor of thinking the operation was real? In part, this may have been because of specific clues that the patients thought they picked up, such as the pain connected with cutting the dura, that both Emily Mason and Jacqueline Winterkorn thought they sensed.

More generally, it's clear that wishful thinking was at work. All the patients very much hoped to have the transplant as their initial surgery, and the rich sensory experience of the surgery—or our simulation of it—made it easy for patients to believe that their hopes had been fulfilled. So they did believe it—even those who, like Jack Celnik and James Ross and Jacqueline Winterkorn, had the scientific training that should have kept them more open-minded. As Jack Celnik put it recently, "For my own psychological well-being, I had to believe that it was the real thing."

The one case where there was skepticism about the operation was that of Sid Howard. It wasn't so much Sid himself who thought he had a sham, but his family. Nee Howard and her daughter were waiting outside the operating room during the procedure, and when Bob Breeze and I came out, they thought they noticed something telling in our demeanor. "You looked too 'loose,' " Nee said recently, "as if you had done something rather trivial. We were both convinced that you had done a sham." It may be that their pessimism spread to Sid himself and countered any impressions he got during the surgery. At any event, he said to Bob Breeze, "You're not going to get another shot at me." He meant that if the surgery was indeed a sham, he would be too old and infirm for a makeup operation a year later. Bob gave him an inscrutable smile and said nothing.

Although most of the patients thought they had received a real transplant, there was enough doubt left in most of their minds that they tried, more or less good-humoredly, to weasel the truth out of Bob or me during the day or two after surgery. They would say things like: "Oh, and there's one more thing that we'd like to ask you . . ." followed by a meaningful look. We were able to counter these little probes with some equally good-humored but uninformative response. Still, it was enough to make me glad that the patients left Denver within a few days to be followed up by Dr. Fahn and his colleagues, who were as ignorant of the truth as the patients themselves.

In setting up the double-blind study in the way we did, we may have stacked the deck against our chances of getting a positive result from it. That's because we made it possible for the patients to experience an extremely strong placebo effect. To see this, consider an alternative way we might have done the surgery. Suppose we had simply put all the patients under general anesthesia for a couple of hours, made small incisions in everyone's foreheads, given half of them transplants, slapped Band-Aids on the foreheads of the others, and then woke them all up and said, "There's a fifty-fifty chance you just had a transplant." In that case, none of the patients would have experienced the drama of the transplant surgery or the equally dramatic pantomimes we put on for the sham surgeries. It seems likely that the patients would have been left in much greater doubt as to whether they had had a transplant or not, and therefore much less likely to get a boost in their subjective sense of improvement. This in turn would have left more room for the transplants to have a beneficial effect over and above any placebo effect.

Indeed, making holes in people's skulls—trephination, as it is called—is one of the most powerful placebo-effect generators known. Trephination as a therapeutic procedure has been practiced by medicine men and faith healers since the Stone Age. When we add the injection (or simulated injection) of fetal cells, we have ratcheted up the placebo effect even further. That, of course, is why one had to be so cautious in interpreting the results of the earlier fetal-cell transplants, and why a double-blind study was so necessary. When we realized that most of our patients were convinced right after the surgery that they had had real transplants, we braced ourselves for a whopping placebo effect and a general improvement in our patients' condition, regardless of whether they had the real operation or not. We could only hope that a difference in outcome between the two groups would emerge as the weeks and months passed by, as the memories of surgery faded, and the neurological benefits of the transplants began to show themselves.

CHAPTER 16

Watching and Waiting

Emily Mason had a headache for a day or two after her surgery. Aside from that, her postoperative recovery was uneventful. On the day after the surgery she was discharged from the hospital, but she and her partner, Giles, had to stay in the Denver area for a while so she could have her stitches removed and be examined for possible complications.

Always a lover of outdoor activities, Mason managed a short hike on the third day after the surgery. A week later she and Giles hiked a few miles in the mountains, near the Continental Divide at the 11,000-foot Hoosier Pass. It gave her a great sense of accomplishment to do something physical that she had not been able to do for some time.

After she and Giles returned home to Toronto, Mason's buoyant mood continued. As the weeks went by, she felt that she was improving both in her physical state and in her ability to think clearly. She began to consider returning to teaching, and Giles had to remind her that she was still quite impaired.

By two months after surgery, things were really looking up. She had one particularly good week in the middle of July 1995. "I was smooth, I was steady, I was functional most of the day, I could walk," she says. "Both Giles and I thought, 'This is it!' " Mason began to tell her relatives and friends of her conviction that she had had the transplant and that the transplant was alleviating her symptoms. They were overjoyed at the thought she was returning to a more normal existence.

Four months after her surgery, Mason returned to New York for her first postoperative evaluation. She went through all the same tests that she had done before the surgery, except for the PET scan, which was to wait until a full year had elapsed. On the Global Rating Scale, the patient's self-assessment of the effects of the surgery, Mason checked "plus 3—greatly improved."

Not long after that first checkup, Mason began to develop debilitating gynecological problems that were eventually traced to fibroids (common, non-cancerous tumors of the uterus). The gains she had noted since her operation were lost again, and she was more affected by dyskinesias that she had been previously. Now she felt it increasingly difficult to live up to people's expectations for her recovery. "People would say: 'You look great,' " she told us recently, "but I looked terrible, I looked like death warmed over."

During this period I had a phone conversation with Mason's Canadian neurologist, who happens to be an expert in movement disorders. He said, "Emily Mason thinks she's wonderful, but I don't think anything's happened to her." Of course, I didn't tell him whether Mason had had a real transplant or not, I just said, "Okay, thanks." or something like it.

In the spring of 1996 Mason underwent a hysterectomy, which put an end to her fibroids problem. Because the operation required hospitalization it counted as a "serious adverse event," and we had to report it to the Performance and Safety Monitoring Board. Not that her fibroids were caused by the surgery we had done on her— most likely they were completely unrelated. It was just a statistic the board kept track of.

The hysterectomy helped Mason regain her color and energy, but her Parkinson's disease stayed about the same. Once so confident that she had had a real transplant, she now began to have her doubts. When she gave a presentation to a Parkinson's group shortly after her hysterectomy, she expressed herself as follows: "At this point, I am not sure whether I have had a transplant. My progress has been overshadowed by gynecological problems. There has been significant improvement, but it has been sporadic and discontinuous over the year."

As Mason's one-year anniversary approached, a definite change in her attitude showed itself—from hoping and believing that she had had a real transplant, to hoping that she had not had one. "I was beginning to realize that if I'd had it, I may have had a little bit of benefit but it wasn't a big deal," she says. She still had severe "off" episodes with freezing, inability to walk, and painful cramps. Eating was still difficult; if she was "on" her dyskinesia ensured that the food went everywhere except into her mouth. If she was "off" the effort required to eat was overwhelming. So she had lost a lot of weight. Clearly, if she had had the transplant, it was not going to revolutionize her life. If she had *not* had it, on the other hand, then the opportunity for a real transplant, with all its hoped-for benefits, still lay ahead.

In May of 1997 Mason returned to New York for her one-year checkup—her third since the operation. As part of the checkup, she had a repeat PET scan. The checkup did not show any obvious neurological improvement in her condition compared with her state before the operation. Once more, Mason was asked to check the box in the Global Rating Scale that corresponded to her self-assessment of her condition since the surgery. She now checked "plus 1—slightly improved." After the checkup, Mason returned to Toronto to await the "unblinding"—the official word as to whether she had had a real transplant or a sham.

Like Emily Mason, Gregory Bennett suffered no serious ill effects from his surgery, which he had in September of 1996. And like Mason, Bennett took advantage of his stay in Colorado to do some outdoor recreation: within two or three days of the surgery, he and his wife, Pauline, went white-water rafting on the Colorado River.

Unlike Mason, Bennett did not experience any subjective improvement during the weeks after the operation. Because nothing changed in Gregory's condition, both Gregory and Pauline rapidly concluded that he had had a sham operation and remained fixed in this opinion throughout the twelve-month follow-up period. The checkups in New York also failed to show any improvement in Gregory's neurological status. If anything, his dyskinesias seemed to be worsening. Since Pauline in particular had never liked the double-blind arrangement, the study now began to seem like a real imposition. They just wanted to get the year over with, receive the official word that the surgery was a sham, and get a real transplant as soon as possible.

Sid Howard had his surgery in February 1996. Since he was our oldest patient—he was seventy-six at the time—we were a little concerned about how well he would tolerate the operation, but he did well. As I mentioned in the last chapter, Sid and his family believed initially that they had a sham operation, based on their observations of Bob Breeze's and my demeanors immediately afterward. Later they became more open-minded about it. Nee noticed what seemed to be an improvement in Sid's speech. Other family members who spoke with Sid on the telephone also commented that he was becoming easier to understand. In addition, Nee observed an apparent improvement in Sid's upper-body strength.

In December 1996, however, Sid was struck by a "serious adverse event." He had traveled from his home in Southern California to Las Vegas, partly to visit his niece and nephew, but also to gamble. One night he was awakened by chest pain, and he was taken to the hospital, where a heart attack was diagnosed. Instead of staying in the Las Vegas hospital, he insisted on returning to San Diego, which he did by himself with considerable difficulty. Once there, he checked into another hospital, where he had an angioplasty to relieve a blockage of one of his coronary arteries. As with Mason's hysterectomy, Howard's heart attack was reported to the Performance and Safety Monitoring Board.

Howard had a rocky time after his heart attack and heart surgery. His doctors advised him to cut down on his fat intake, but somehow he took that as a recommendation to eat less food, and within a few months he had slimmed down from 165 pounds to a skin-and-bones 112 pounds. At that point his doctors said, "Eat junk food, eat anything, just get your weight back up!" Howard did regain some of his former weight, but not all of his energy.

What with his Parkinson's disease and his heart problem, Howard found the trips to New York for checkups exhausting and asked to be excused from making them. Not wanting to lose him from the study, Stan Fahn actually made a transcontinental house call and did the assessment in Howard's home. On a subsequent occasion Sandra Dillon, the nurse who was responsible for Howard's testing, did the same thing. Thus we managed to get all the crucial information. At twelve months after surgery, Howard's Unified Parkinson's Disease Rating Scale score was modestly improved.

In spite of Howard's heart problems, both Sid and Nee felt sure that there had been some improvement in his condition by the time the anniversary of his surgery arrived. They had completely changed their views from what they had thought right after the surgery, and they were expecting to hear he had had a real transplant.

Jack Celnik went through the opposite experience to Sid Howard. Right after his operation, which took place in January 1997, Celnik believed he had had a real transplant, but more for internal psychological reasons than because of any actual evidence one way or the other. Celnik did not notice any improvement during the following weeks, however, and as the months rolled by he sensed that his disease was actually progressing. "By three to four months after the operation I had a pretty firm sense that it was a sham," he says. After that, he looked forward to being told that it was a sham and to the opportunity to receive a real transplant.

James Ross's surgery was in August 1997. Like Celnik, Ross believed right away that he had had a real transplant. For the first week after the surgery, his condition was worse than before, but then he returned to his preoperative state. Over the course of the following three months Ross and his wife, Gail, noticed what they thought was some modest improvement, especially in his ease of movement while "off."

In early December 1997, shortly after returning to San Diego from his four-month checkup in New York, Ross's apparent progress was set back by an unfortunate episode. One night he fell on his way to the bathroom. When Gail asked him whether he was all right, he tried to reply, but his words were garbled. Fearing that he was suffering a stroke, Gail called an ambulance. Before he reached the hospital his speech and movement had returned to normal, so the episode may have been a "transient ischemic attack"—a temporary interruption of blood flow to part of the brain—rather than a stroke proper.

Whether it was a stroke or a TIA, the cause of episode needed to be investigated. An ultrasound examination showed that Ross's right carotid artery was largely blocked by atherosclerosis. Ross had to undergo an operation to clear the blockage. The operation was successful, and Ross suffered no further ischemic episodes, but the whole experience did cause his general level of well-being to deteriorate somewhat. And it was another "serious adverse event" for the PSMB to review.

By the end of the twelve-month postoperative period, Ross's condition was not greatly changed from his preoperative condition, whether by his own assessment or by the testing at Columbia. Ross oscillated in his opinion as to whether the operation had been real or a sham. A part of him hoped the operation had been real, and another part of him hoped it had been a sham, so that he could look forward to more significant benefit from the makeup surgery.

After her surgery in January of 1998, Jacqueline Winterkorn felt optimistic that she had had a real transplant. This was partly the general wishful-thinking optimism that affected most of our patients, and partly the consequence of the episode during surgery when she thought she felt the brain membranes being cut. At any rate, she was too much of a scientist to put all her money on one interpretation. "I tried to keep an open mind," she says.

The only complication Winterkorn had from the surgery was a skin infection. In making the incision in her forehead, Bob Breeze had cut through a small cyst that Winterkorn had had since childhood, and the cyst became infected. A few applications of antibiotic ointment quickly cleared it up.

During the first week after the operation, Winterkorn felt that she was more "brittle" than usual—that is, she experienced more frequent on-off fluctuations. She wasn't sure whether that was the result of the surgery or simply the effect of having to go without her usual medications for many hours. (All the patients had to stop medications on the morning of surgery.)

As soon as she returned home to Connecticut, Winterkorn went back to her usual work routine—the long commute to her practice on Long Island and her teaching responsibilities at New York Hospital. For a few weeks, her condition remained unchanged. By three or four months after surgery, however, she felt that she was slightly but distinctly improved. There was no change in the severity of her dyskinesias, but her "off" periods were becoming less frequent, shorter, and less severe.

Winterkorn had now been bereaved for nearly a year, and she had had to rely largely on herself to get through the operation and to cope with her continuing disability. In May 1998, however, her personal life took a turn for the better. "I had gone on the Internet one night to check some references," she says. "Somehow I found myself in one of those 'find-a-person' programs, and I tried to figure out someone I wanted to find. I typed in the name of my first boyfriend, David Lincicome, whom I met in summer camp when I was fourteen years old. I found him in Ohio, and we struck up an e-mail correspondence. First it was once a day, then three times a day, and then we were doing 'instant messages' in real time."

Within two weeks, David wanted to come to Connecticut to visit Winterkorn, but she was dubious. "I said, 'Don't do that, I'm not fourteen anymore, I'm fifty, I have Parkinson's disease and I look terrible, I'm an old person. You don't want to have a relationship with me, let's just continue to correspond.' " But in June he did visit, and later that summer he moved in, after transferring his law license to Connecticut. A year later they were married.

David was accepting of Winterkorn's illness—he said that her dyskinesias were "cute"—and he helped her deal with many issues, medical and otherwise. But Winterkorn feels that it was the improvement in her condition that motivated her to make that initial contact and allowed her to reciprocate David's interest. "I don't think I would have felt comfortable committing to a relationship unless I felt that I was getting better," she says.

I was left wondering whether our protocols should have kept track of "happy events," which might have counterbalanced the adverse events that some of our patients experienced.

By six months after her surgery, Winterkorn was convinced she had improved. Before her surgery she had awakened in the morning in an "off" state and had to take her medication and wait for forty-five minutes before she could get up and take a shower without falling. But now she was waking in an "on" state and she could shower and brush her teeth even before her first morning dose took effect. Throughout the course of the day, her movements were brisker and more precise. Still, her dyskinesias were not particularly improved.

As the remainder of her first postoperative year went by, things continued to get better. Both her gait and her speech improved. Also, she was able to reduce her L-dopa dosage by one-third—from 1.2 grams daily to 0.8 grams. With the reduction, she experienced a lessening of her dyskinesias. The checkup at twelve months confirmed Winterkorn's own sense that she had improved markedly. Thus, Winterkorn was so convinced that she had had the real implant that her upcoming anniversary and official "unblinding" was not a matter of great concern to her. She just concentrated on her work, on her new relationship, and on taking advantage of her improved condition to enjoy life generally.

CHAPTER 17

Inside the Brain

Traveling in the morning from New York City to the North Shore University Hospital at Manhasset, on Long Island, is like swimming against the tide. Dodging the waves of disembarking commuters at Penn Station, one clambers aboard a near-empty train as it heads back eastward to pick up another load. The journey is also a trip backwards in time, from Manhattan's frenetic modernity to communities that seem frozen in the 1950s. Gazing from the train window at the tidy rows of suburban homes, one expects to see Avon ladies or Fuller Brush men hawking their wares from door to door.

Although the North Shore Hospital is primarily a community hospital it also has an academic slant, thanks to its affiliation with New York University Medical School. Among its facilities available for clinical research is a positron-emission tomographic camera, or PET scanner; all our patients have had to put their heads inside this machine on more than one occasion during the study. The PET-scanning technique has been able to tell us, better than any other available method, the fate of the transplanted tissue inside our patients' brains.

The PET technique allows one to "see" human biochemistry—the fundamental chemical processes that make our existence possible—from outside the body. To do this, it detects and measures the distribution of ordinary bodily chemicals that have been tagged with radioactive atoms. Specifically, the tags are radioisotopes that when they disintegrate emit positrons. Positrons are the antimatter versions of electrons—they are like electrons except that they carry a positive charge rather than the usual negative charge of an electron.

The advantages of using positron-emitting radioisotopes are two-fold. First, they are short-lived: most of them undergo radioactive disintegration within a few minutes or hours of their creation. In the process of disintegration they change into harmless, nonradioactive atoms. Thus a patient can be given high doses of these radioisotopes for the purpose of a PET scan, and within a few hours most of the radioactivity will have disappeared. This keeps the patient's total radiation exposure within safe limits. The other useful feature of positron-emitting radioisotopes has to do with the peculiar manner in which they disintegrate, which I'll describe later.

In cases of patients with Parkinson's disease, we want to observe dopamine cells, and we therefore use a radioactively tagged version of L-dopa. One of L-dopa's hydrogen atoms is replaced with the positron-emitting radioisotope of fluorine, 18-F. The resulting molecule is called 18-F-L-fluorodopa, or just fluorodopa. Dopamine cells take up the fluorodopa as if it were regular L-dopa, and it is converted to radioactive fluorodopamine by the same enzymes that naturally convert L-dopa to dopamine. The radioactivity therefore accumulates in the cell bodies and synaptic terminals of the dopamine neurons. By imaging the distribution of the fluorodopa, the PET scanner shows us where the dopamine cells are located within the patient's brain.

The isotope 18-F decays with a half-life of 109 minutes; that is, in every 109-minute period one-half of the total 18-F atoms in a sample will have disintegrated. Although this short lifetime is medically advantageous, it is highly inconvenient in a practical sense, because it means that one cannot order it from some distant supplier, nor can one store it for any period of time. At the North

Shore Hospital the 18-F is manufactured on site, with the help of a machine known as a cyclotron.

Housed in a cavernous space in the hospital's basement and shielded by eight-foot-thick concrete walls, the cyclotron is a particle accelerator that generates a beam of high-energy protons. The beam is aimed at a small sample (about ten drops) of water that sits in a small chamber behind a thin metal foil. This is a special form of water, however, in which the oxygen atoms are the 18-O isotope rather than the common 16-O isotope; that is, their nuclei contain eight protons (thus defining the atoms as being oxygen) but they are accompanied by ten rather than by the usual eight neutrons. When a sufficiently energetic proton from the beam strikes an 18-O atom, it merges with the nuclear particles, knocking out a neutron in the process. The resulting atom has nine protons and nine neutrons, so it has become a different element, fluorine 18 (18-F).

When a sufficiently large quantity of 18-F has been created, the radiochemist turns off the cyclotron, so that it is safe to enter the room. The chamber now contains a small quantity of 18-F in the form of fluorine gas. He takes this sample to a heavily shielded chemical hood, where it is used to synthesize fluorodopa. The chemist has to work fast, because the 18-F is decaying with a half-life of 109 minutes. Actually, 18-F is a little more accommodating in this respect than other positron-emitting isotopes, some of which have half-lives as short as a couple of minutes.

Once the fluorodopa has been synthesized, it is brought to the patient and injected into a vein under the supervision of Vijay Dhawan. The fluorodopa passes through the bloodstream to the brain and is taken up by dopamine cells. The task for the scanner is now to generate an image of where the fluorodopa is located.

This is where the peculiar manner in which positron-emitting radioisotopes disintegrate comes into play. The unstable 18-F nucleus wants to relieve itself of its excess positive charge, which it does by converting one proton into a neutron. A positron and a neutrino are emitted in the process, and the atom now reverts to its original 18-O form. The neutrino takes off at a high rate of speed and is never heard from again, but the positron has a more interesting career, though a brief one. Like a ball in a pinball machine, it strikes and bounces off a number of nearby particles, losing kinetic energy in the process. Finally, after a tiny fraction of a second, it has lost enough energy that it can do the thing that all antimatter wants to do in this world, which is to annihilate itself in a collision with its matter equivalent—an electron in this case. Two gamma rays (high-energy photons) carry off the energy released in the annihilation, and in order to conserve the total momentum of the system they have to fly off in opposite directions from each other. Most of the gamma rays generated deep within the patient's brain are able to penetrate the brain tissue and escape from the head, and they then continue outward at the speed of light.

The PET scanner, which like the cyclotron costs about $2 million, is an assembly of gamma-ray detectors. In the North Shore Hospital machine there are about 12,000 of them, forming a cylindrical array about three feet in diameter and six inches long. The patient's head is placed within this array, so a gamma ray emerging from the head is likely to strike one of the detectors, triggering an electronic signal.

The detection of the gamma rays is only the beginning of the game, however; the machine then has to figure out exactly where the gamma rays are coming from. The first step in this procedure involves coincidence detection: if two detectors on opposite sides of the cylinder receive a gamma ray almost simultaneously, the scanner assumes they arose from the disintegration of a single positron. Therefore, the disintegration must have occurred somewhere along the straight line between those two detectors. The machine cannot tell where along that line that disintegration occurred. However, we are dealing not with a single disintegration but with enormous numbers of disintegrations that occur during the time the patient is lying in the scanner. Using appropriate statistical procedures the machine figures out the most probable distribution of radioactive atoms that could have led to the observed pattern of gamma-ray detections. This corresponds closely to the actual three-dimensional distribution of fluorodopa in the patient's brain.

The machine then displays a series of slices through the brain, starting at the top of the cerebral cortex and working down to the brainstem, with the fluorodopa distribution mapped in color on each slice. The map has a certain fuzziness to it; it does not show single dopamine cells, for example, only the general regions in which these cells are located. One reason for this is that positrons travel a few millimeters from their source, in some random direction, before they annihilate and give rise to the gamma-ray pair. Still, the map is sharp enough to show differences in fluorodopa uptake in different subregions of the striatum.

For a PET scan, patients are asked to discontinue medications the previous evening, so by the time of the scan they may be feeling like an untreated Parkinson's patient, which is to say slow and stiff, perhaps with some unwanted movements. The fluorodopa is injected into a vein, and they are allowed to relax for forty minutes while the drug enters the brain. Then they are positioned in the scanner, with their head immobilized. Luckily, PET scanners are not as claustrophobic as some other machines, such as MRI scanners, because the PET scanner's tunnel is so short.

The main problem is the necessity for keeping the head immobilized for about 75 minutes: 10 minutes for a calibration scan, 5 minutes for a two-dimensional scan, and then 60 minutes for the full three-dimensional scan. This can be tough on a healthy person, but it is extra demanding for some patients with Parkinson's disease. "The PET scan is very difficult," Emily Mason tells us, "because it is so hard to lie still: I'm very restless, and eighty minutes is a long time. My neck can hardly tolerate it. Giles has to read to me, and I'm always asking, 'How much time do I have left?' " Jacqueline Winterkorn, who had sent many patients to be scanned in that very machine when she was working at North Shore Hospital, says simply, "I'd much rather be the doctor than the patient."

Trying though it is for the patients, the information derived from the PET scans has been vital to the study. In scans performed during the patients' initial evaluations, the data revealed the severity of their Parkinson's disease: the striatum "lit up" much less than it should in a normal patient, meaning that there were fewer dopamine terminals (the nerve endings of the nigro-striatal pathway) available to take up the fluorodopa. The loss was generally worse in the division of the striatum known as the putamen than in the other division, the caudate nucleus, and that is why we made our fetal-cell injections only into the putamen.

David Eidelberg, the neurologist who runs the Parkinson's disease unit at the North Shore University Hospital, used the PET scans to get a quantitative measure of the severity of the loss of dopamine terminals in each patient. Basically, he compared the PET signal in the patient's striatum, where dopamine terminals should be abundant, to the signal in another "control" region of the brain where there should be few or no dopamine terminals. In a healthy person the ratio of these numbers is about 2.3—that is, the PET signal is about 2.3 times as strong in the striatum as in the control region. The ratio can be as low as about 1.8 without causing any problems; below that, some symptoms of Parkinson's disease are likely. In the patients enrolled in the double-blind study, the ratio before surgery ranged from about 1.4 to 1.8. These were the baseline figures with which the patients' postoperative scans could be compared.

All the patients had a second PET scan twelve months after surgery, shortly before they were "unblinded" and told whether or not they had received a real transplant. Not surprisingly, the results of these second scans were quite diverse. In some, the ratios calculated by Eidelberg increased toward normal levels. In Sid Howard's case, for example, his initial ratio was about 1.6, down in the "pathological" range, but by a year after his surgery the ratio had gone up to 2.0, which is in the "low normal" range. In Jacqueline Winterkorn's case the ratio also rose over the first year, but more modestly: from about 1.4 to about 1.6. In some other patients the ratios stayed pretty much unchanged over the twelve-month period; in a few others it actually decreased.

Like all the staff at the North Shore facility, Eidelberg was kept in the dark about who had the real transplants and who the shams. For that very reason, he could make a key judgment call: based on the PET scans, had each patient received a transplant or not? This was a critical part of the data analysis. While a computer calculation could give a number that could be used to compare two scans, a human observer is best able to say whether something has changed between the preoperative and postoperative images. Of course, wishful thinking could create a bias, so blinded ratings by Eidelberg would show how detectable the transplants really were. Seeing the PET data he thought that Howard must have received a real transplant, for example, and he felt that Winterkorn did too. Eidelberg wrote down his best guesses as to which patients fell into which groups. He and his staff also calculated the quantitative ratios for each patient. All of this was done blindly for each patient, and the results were submitted to our data collection center before the code was broken for that patient. By noting how often Eidelberg and his team had been right, we would get some measure of how well the grafts had survived and how well the PET technique had been able to detect their presence.

Valuable though the PET technology has proved to be, it is far from telling us all we want to know about the fate of the transplanted cells. By itself, increased fluorodopa uptake might be due to a variety of factors. For example, it's conceivable that the transplant surgery simply damages the patient's striatum in a nonspecific way, and this damage then causes sprouting of new terminals from the patient's own nigrostriatal pathway. These new terminals could then take up extra fluorodopa and cause a brightening of the PET image, even if all the transplanted cells had died. And assuming that the increased uptake is indeed due to the presence of the transplanted cells, as we hope, there still remains much uncertainty as to the interpretation of the findings. Because of its limited resolution, for example, PET scanning can't distinguish between dopamine cell bodies—which is what we transplant—and dopamine terminals, which are the business ends of the fibers put out by dopamine cells. In one scenario, the transplanted cell bodies might just sit there and do nothing. In that case they might help light up the striatum in the PET image but would not actually alleviate the patient's symptoms in any way. In another scenario, the cells might grow out thousands of fine fibers, and the terminals of these fibers might interact with the native architecture of the striatum, providing a functional replacement for the missing terminals of the nigrostriatal pathway. We know that this is what happens in animal experiments, but does it happen in humans? PET scanning alone cannot give us the answer.

The most definitive method to analyze what is happening inside the brains of transplant recipients remains the same as it was in James Parkinson's day—autopsy. Much as we would like all our patients to survive indefinitely, the fact is that some do die from a variety of causes, and the study of these patients' brains offers insights into the fate of the transplanted tissue that can be obtained by no other procedure.

As mentioned in chapter 9, one such case was reported by Gene Redmond and his Yale colleagues. This patient, who had received previously frozen tissue from an eleven-week-old fetus, died four months after the transplant operation. Only a single dopamine cell could be found in his striatum. The poor result was probably due to both the freezing of the tissue and the age of the fetus: the optimal fetal age for transplantation is about seven weeks, and eleven weeks is borderline in this respect. In chapter 11 I mentioned another case, of Max Truex, who received a transplant from Robert Iacono in China. None of the nerve cells transplanted into his striatum were still present when his brain was autopsied two years later. A number of factors probably contributed to their failure to survive, but the most important was probably the age of the fetuses from which they were derived—sixteen weeks. This is far outside the acceptable range.

In 1995 Warren Olanow's group reported on another case. This was a fifty-nine-year-old man who had received bilateral grafts of tissue from fetuses aged between 6.5 and 9 weeks. He showed remarkable improvement after the transplant: by three months both his dyskinesias and his "off" episodes had essentially disappeared, and this improvement persisted over the next year. Consistent with this improvement, PET scans done at six and twelve months showed a progressive increase in fluorodopa uptake in the putamen, compared with his preoperative scan.

The patient had broken his ankle many years previously and to alleviate his persistent pain an orthopedic surgeon fused his ankle joint eighteen months after the transplant. Unfortunately, he died of a massive pulmonary embolism while recovering from the ankle surgery. Most probably, a blood clot formed in one of his leg veins, broke off, and was carried in the bloodstream to the lungs, where it blocked the pulmonary artery or one of its major branches.

Because the patient had died while still in hospital, Olanow was able to arrange for an autopsy to be performed just four hours after his death, and his brain was placed immediately into chilled fixative. Thus the preservation of the brain tissue was optimal. The brain was shipped for analysis to Olanow's collaborator, Jeffrey Kordower, a neuroscientist at Rush-Presbyterian–St. Luke's Medical Center in Chicago. Just as Judith Folkerth had done with the brain of Max Truex, Kordower used immunohistochemistry to visualize tyrosine hydroxylase, the enzyme that is found in dopamine cells. With this labeling technique, they found dense clusters of healthy-looking dopamine neurons at all the injection sites. In fact, a total of about 200,000 transplanted dopamine cells had survived. This number is about 50 percent of the normal complement of dopamine cells present in a healthy human brain. Since people can lose about 80 percent of their dopamine cells before they show symptoms of Parkinson's disease, a 50 percent replacement obviously has great reparative potential. What is more, fibers from the transplanted cells had grown out into the host tissue, forming a network of fibers throughout much of the putamen, just as had been observed in animal experiments.

In September 1996, one of the patients in our double-blind study, a sixty-six-year-old woman from South Carolina named Mildred Timmons, died in a traffic accident. Timmons had had her surgery seven months previously, and it happened to be a real transplant, not a sham. Like the other patients, she had undergone an evaluation in New York four months after her surgery, and the results were very positive—she showed a 30 to 40 percent improvement in the aggregate results of her tests. Three months later, she was driving her car near her home while the eye of Hurricane Fran was charging through North Carolina, leaving devastation in its wake. Although South Carolina was on the fringes of the hurricane, it too took a tremendous beating—up to thirty inches of rain fell, and trees were downed everywhere. One tree fell directly into Timmons' path, and in swerving to avoid it she lost control of her car, drove into a ditch, and was killed.

To lose a patient is painful under any circumstances, but for a patient in our study to be one of the thirty-four people who died in Hurricane Fran seemed like the most diabolical stroke of ill fortune. Recall that of the forty participants in our study only twenty patients were in the experimental group—the group who received real transplants at their first surgery. And in the older group of patients, to which Timmons belonged, there were only ten people in the experimental group. Losing one of those ten—one who seemed to be doing well—before she even reached the one-year mark was a major blow.

Mildred Timmons's family was in shock at this tragedy, of course, but they knew we would want to examine her brain, so they called Fahn's unit at Columbia to let them know what had happened. The clinical coordinator was able to locate a pathologist willing to collect Timmons's brain in spite of the ongoing storm. All electrical power to the area had been cut off, and it was after nightfall. The pathologist had to find his way to the funeral home in terrible weather and in complete darkness, and then he had to remove Timmons' brain with only a flashlight for illumination. Since there was no refrigeration, he had to place the brain in warm formalin. That, and the rather long time that had passed between Timmons's death and the autopsy, meant that the tissue was not ideally preserved for the immunohistochemical procedure.

When we received the brain in Denver, we wanted to slice it in the same plane as made by the needle tracks during the transplant surgery: this way we would be able to see each entire track, and the row of graft sites along that track, in a single slice. There were no particular landmarks on the surface of the brain that we could use to get the right orientation, however. So we decided to do an

MRI scan of the brain before slicing it. Although it's unusual to do an MRI scan on an isolated organ after the patient's death, the procedure actually worked very well, giving us an excellent three-dimensional view of the inside of the brain.

We expected to see the needle tracks as four obvious streaks of damaged tissue running nearly horizontally from the front of Timmons's brain back to the striatum. However, we couldn't see any tracks at all, in spite of the high quality of the image. This gave us pause. Indeed, we began to suspect that this was the wrong person's brain. Could it be that in the confusion and darkness of the hurricane the pathologist in South Carolina had mistaken the identity of the person whose brain he had removed? Fearing the worst, we went back to Timmons's records and pulled out the MRI scan she had had at the time of her original surgery. We compared the folding pattern of the cerebral cortex in the two scans. This pattern is as unique to an individual as his or her fingerprints, and we found that the patterns in the two scans were identical. Thus we were able to convince ourselves that we did indeed have Mildred Timmons's brain, and not the brain of some other unfortunate individual whose body happened to be lying in the funeral home on the same evening.

We now reexamined the postmortem MRI scan in minute detail. The scan is stored as a large computer data file. We played the scan slice by slice, moving through the brain like a slow-motion movie. Finally we saw one tiny dot that moved from image to image like a faint star moving across the sky. This had to be one of the tracks. Using this as our guide, we were able to orient the brain in the correct plane for slicing.

Having obtained our slices, we stained some of them to see the general layout of the tissue. Under the microscope, we could now identify all four needle tracks. The reason we had had such a hard time finding them in the MRI scan was that the needle had caused very little damage on its four passes into the brain. This was probably on account of the extra-fine diameter of the needles we were using at the time of the double-blind study.

To identify the grafted dopamine cells we used the same procedure as Olanow's group had employed—immunohistochemistry.

This is a multistep procedure that takes advantage of the amazing specificity and sensitivity of the immune system to visualize particular kinds of molecules present in tissue slices. It can detect molecules that are present in extremely low quantities, as is true for tyrosine hydroxylase, the enzyme present in dopamine cells.

In simplified outline, the procedure works like this. In the first step, we bathe the tissue slices in a solution that contains a few drops of serum from a rabbit that has been immunized against tyrosine hydroxylase. The antibodies in the rabbit's serum lock onto the tyrosine hydroxylase molecules in the tissue. Then we treat the slices with serum from a goat that has been immunized against rabbit antibodies. The goat antibodies have been previously linked to an enzyme that can produce brown-colored deposits when exposed to certain chemicals. These specially treated goat antibodies lock onto the rabbit antibodies that are in turn attached to the tyrosine hydroxylase molecules. Finally, the slices are exposed to the chemicals that trigger the brown reaction. Because the enzyme capable of forming the brown deposits is linked to the goat antibodies which are in turn linked to the rabbit antibodies which in turn are linked to the tyrosine hydroxylase molecules, the brown deposits form only where tyrosine hydroxylase is present in the tissue—which is to say, in the dopamine cells.

Complicated? Yes, but also exquisitely sensitive, because each step of the procedure amplifies the effect of the one before it. In fact, despite the less-than-optimal conditions in which the brain was obtained and preserved, we obtained beautiful staining of the dopamine cells in Timmons's striatum. As we had hoped, they were present in abundant numbers along all four tracks, clearly marking where we had laid down the four "noodles" of fetal tissue. In addition, the cells had grown out fibers that invaded the brain tissue around the tracks, just as we had hoped, although they did not yet form a confluent network.

One other person in our double-blind study died. His name was Eugene Weiner, and like Timmons he was in the older group and received a real transplant. He was sixty-eight at the time of the surgery. A year later, his PET scan showed a marked increase in fluorodopa uptake in the striatum on both sides of the brain, and his scan at two years after the transplant was just as good. Weiner did not show any clinical improvement one year after his transplant, but by three years his UPDRS scores were significantly improved. This delay in improvement has characterized our older patient group.

Weiner had had high blood pressure, and not long after his three-year evaluation he had a heart attack and died at age seventy-one. His wife agreed to allow us to study his brain. As with Timmons, the tyrosine hydroxylase immunohistochemistry revealed abundant surviving dopamine cells along the needle tracks, although the track with the most cells had almost ten times as many cells as the track with the fewest cells, showing the variability of transplant cell survival. There were two other remarkable things about Weiner's brain. First was that the dopamine fibers growing from the cells had formed a complete network throughout the putamen, overlapping the individual transplant tracks. This result showed that three years was long enough to get a complete resupply of the dopamine network. Since the beginning of our work, we have seen progressive improvement in patients for about three years. Now we had the anatomic proof that growth of the transplant was complete by three years, with a much greater network than seen at seven months in Timmons.

The other significant finding was that the individual dopamine neurons in each of his four transplants had begun to accumulate the black pigment granules typical of adult human dopamine neurons. As mentioned earlier in the book, it is the pigment in dopamine neurons that give the brain region containing them its name, substantia nigra ("black substance"). Children are born without any pigment in their dopamine cells, but it accumulates over the first decade of life. The fact that pigment was developing in the dopamine neurons in Weiner's brain showed that the transplanted cells were developing on a normal timetable.

Since then, three more of our transplant patients have died, up to eight years after transplant. These were patients that we operated on prior to the double-blind study, and all three showed excellent survival of the transplanted cells. Remarkably, the pigment granules in the dopamine neurons were larger and darker than those in Weiner's brain, further indicating that the implanted dopamine neurons have a normal development pattern over at least eight years after transplant.

Thus the overall autopsy results show that transplanted fetal dopamine neurons can survive for years after transplant and can reinnervate the dopamine-deficient host brain. They also strongly support the assumption that an improved PET scan (increased fluorodopa uptake in the striatum) in a transplant recipient is caused by the survival and growth of the transplanted dopamine cells and their fibers, and not by some nonspecific effect of the transplant surgery.

On the other hand, Weiner's case illustrated something else that we were less happy to acknowledge. Weiner's PET scan was markedly improved at one year after his transplant, but he showed no clinical improvement at that time. Evidently there was no direct correlation between improvement in the PET results and improvement in the patient's clinical condition. As January 1999 approached— the date when Jacqueline Winterkorn, our last patient, would reach her one-year anniversary and be "unblinded"—we looked forward with a mixture of excitement and anxiety to the opportunity to sit down and analyze all our data. We would finally know what the real benefits, if any, of the fetal-cell transplant surgery might be.

CHAPTER 18

Anniversaries

When May 1996 arrived, and Emily Mason had completed her one-year postoperative period, she looked forward with a mixture of anxiety and hope to finally learning whether or not she had a real transplant. Anxiety lest the answer was yes, in which case the procedure had brought her little benefit, and hope that the answer was no, in which case she still had the option of a real transplant ahead of her. Yet, as it turned out, Mason did not get any answer in that month, or even in that year.

A portion of this delay had been anticipated. Before the blind could be broken, Mason had to have her one-year checkup and PET scan and the data from those tests entered in the study's database. But that should have taken no more than four weeks. As August came around and Mason was still none the wiser about her surgery, she grew understandably impatient. She faxed a letter to Fahn and myself expressing her dissatisfaction. "Much to my dismay," she wrote, "the blind has not been broken. As the days go by, this is making my health and my life practically more and more difficult. . . . The lack of knowledge about whether I will be having surgery again or whether I have the tissue causes upheaval to my life and to those who support me. A short delay of a few weeks was bearable; this present indecision, however, extended over several months, could have a far-reaching negative effect on the results of the study, and certainly on my day-to-day functioning. I am less able to understand the fluctuations of the disease, to plan realistically, and adapt positively."

The reason for the delay was as follows. Around the time that the first anniversaries approached, the Performance and Safety Monitoring Board discussed what was to be done about the "makeup" surgeries for the patients whose initial surgeries turned out to be shams. The members of the board felt that the makeup surgeries should not be done until the results of the entire study were in. After all, only then could the patients make an informed decision as to whether they were likely to benefit from the procedure.

But if the makeup surgeries were to be postponed, the board reasoned, what sense did it make to tell the patients whether or not they had had a real transplant? The patients would be making no decisions based on that information, so why not continue the blind for all patients until the last patient had reached his or her one-year anniversary? Because the initial operations were to be spread over two to three years, this would give us results under double-blind conditions that covered considerably longer than the single year originally planned—up to nearly three years in some cases. Why throw away the chance for this extra windfall of scientific data?

Having reached this conclusion, the PSMB's chairman Ira Shoulson told the NIH not to allow us to begin the makeup surgeries or even to tell patients which group they had been in. The NIH concurred and passed on those instructions to Stan Fahn and myself. We had to tell Mason and a couple of other patients who were in the same position that there was nothing we could do—our hands were tied by the NIH.

None of the patients were pleased with this turn of events—they felt the rules had been changed in the middle of the game. Eventually, the outcry was such as to force the NIH to acknowledge that the patients had a right to disclosure at one year and to makeup surgery if requested. Thus, in February of 1998, Paul Greene called Mason with the fateful news. Reading a prepared statement over the telephone, he told her that her transplant had been a sham and that she had the right to receive a real transplant. Mason was over-joyed, and immediately told Greene that she wanted the transplant.

A couple of months later, in April 1997, Bob Breeze and I per-formed a real transplant on Mason. It was quite a different experi-ence for her from the first time around. For one thing, the experience of having the halo put on was much less stressful, because Bob Breeze had started giving the patients morphine before the procedure. "You don't feel nearly as much pain because you're flying as high as a kite," is how Mason put it.

The period after the transplant, on the other hand, was much more stressful than after the sham surgery. Although Mason felt euphoric for the first twenty-four hours, in fact was able to walk out of the hospital unaided, she went rapidly downhill after that. On the second night after the surgery, while staying at a friend's home in the Denver area, she awoke almost completely unable to move. For her checkup with Bob Breeze the following week, she had to be brought into the hospital in a wheelchair—not only was she unable to walk but her head was lolling uncontrollably to one side. She also complained of difficulty in concentrating.

Bob reassured her that the setback would be temporary, and this turned out to be the case, although her recovery took quite a time. By six to eight weeks after the operation she could hold her head up and her walking and thinking were somewhat improved. After that, things got steadily better. By nine months, she was markedly better than her preoperative state: she could walk on command, her posture improved, and her on-off fluctuations leveled out. She began to cut back on her medications, and by the end of the first year she had cut her daily L-dopa dosage almost in half.

The impact on the quality of Mason's life was enormous. "I was able to leave the house by myself—that was a major change," she says. "And I could sleep through the night. Before, I'd sleep as much during the day as during the night, so I'd be asleep when other people were awake and awake when they were asleep. I was able to move much more smoothly. And my dyskinesias were less. Before, I could hardly have a phone conversation because I'd be flinging the phone around the room. I started my artwork again, and Giles and I were able to go for walks. We could go to the movies again—I remember sitting through *Titanic* and getting up and being able to walk afterward—it was incredible! I went on a business trip with Giles and I went to San Francisco to participate in a dance workshop."

By the time Gregory Bennett's anniversary came around, both he and his wife had long since become convinced that he had had a sham operation. Thus they were not at all surprised when, in February 1997, Paul Greene called and told them that it had indeed been a sham. Like Emily Mason, Gregory Bennett immediately said he wanted a real transplant. Unfortunately, a trail of circumstances made it even harder for Bennett to get his makeup surgery than it was for Mason.

What happened was this. By early 1997 most of our patients had had their initial surgery; only James Ross, Jacqueline Winterkorn, and a few others were still waiting for their operations. Among the patients who had already been operated on, a number of "serious adverse events" had occurred. First and foremost was the death of Mildred Timmons, whose hurricane-related car accident had occurred in September 1996. There was also Emily Mason's hysterectomy in early 1996, Sid Howard's heart attack in December 1996, and four other events: another heart attack, a suicide attempt, a broken toe, and a confusional episode. The confusional episode turned out to be due to a subdural hematoma (a bleed on the surface of the brain), but the reason for the patient's hematoma—whether a late consequence of the surgery or from some other cause such as a fall—was never clear. (An eighth adverse event, James Ross's carotid artery problem, was still in the future.)

At a meeting in mid-1997, the PSMB reviewed these occurrences. The board members realized that there was an imbalance in the distribution of adverse events, with most having occurred among the patients who got the real transplants.

PSMBs have the authority to intervene or even halt a study if patients are being harmed. In this case, the PSMB told us that we could continue to operate on the few remaining participants who had not yet had their initial surgeries, but that we should stop doing makeup operations for the patients whose initial surgeries had been shams. This was just what they had said a year earlier before being overruled by the NIH, but now they had an apparently serious safety issue on their side, so the NIH sided with them. The decision came too late to affect Emily Mason's makeup surgery, but Fahn had to tell Gregory Bennett and several other patients that their makeup surgeries would have to be postponed until all the patients had been unblinded.

The Bennetts were far from happy. It was now over two years since Gregory had enrolled in the study. As several patients had not yet received their initial surgeries, they could be sure that it would be at least another year, and probably more, before the results would be decoded and they would be allowed to decide whether to have the makeup surgery. Pauline Bennett in particular had never been enamored of the double-blind design. She was even less enthusiastic, now that she realized it would take up their time and energy for more than three years before they stood any chance of benefiting.

Pauline was forthright in communicating her opinion to Fahn. "I reminded him that we were supposed to be able to get this surgery right after we were notified that the first surgery was a sham," she says. "I told him that we had a contract and we had signed it legally and everyone else had signed it legally." But Fahn explained that his hands were tied—he could do nothing without a clearance from the PSMB. And he went over the adverse events that had triggered the PSMB's concern.

This information incensed the Bennetts even further. The idea that a traffic accident caused by a hurricane could have any impact on the continuation of the study struck them as ridiculous.

And as for the hysterectomy—"I couldn't see Gregory needing a hysterectomy," says Pauline. Even the heart attacks and the other adverse events seemed to have little direct connection to the transplants. Surely this was all coincidental—the kind of things that just happen to people, especially to seriously ill people who are no longer young.

My own reaction to all this was a little more cautious. In general, PSMBs are justified in taking all adverse events into consideration, not just those that are obviously linked to the study. If they started to pick and choose which adverse events to consider, there would be a serious risk of missing possible harmful effects of the procedure being studied. Take the adverse event that had the least apparent connection to the study, the death of Mildred Timmons in Hurricane Fran. With a bit of effort, one can imagine how the transplant might have contributed to her death. After all, I've already mentioned how quite a few of our transplant recipients, once they noticed an improvement in their condition, developed an irrepressible urge to get behind the wheel of a motor vehicle. Was this why Timmons was out on the road in life-threatening weather? Probably not, but we don't know that for a fact. Gregory Bennett would have chosen to face any and all traffic hazards rather than continue his enslavement to Parkinson's disease. But the PSMB has to look at the big picture.

Regardless of the PSMB's legitimate concerns, both Stan Fahn and I felt that the sham-operated patients had the right to opt for makeup surgery right away, based on their initial agreements. We also thought that the operations had played little if any role in triggering the adverse events—in other words, the imbalance in the numbers was largely a coincidence. But there was not much we could do about it except suggest that the patients who were demanding the makeup surgery let NIH know how they felt.

Another patient who had been told that his initial operation was a sham, a retired New York lawyer, "Charles Frinton," was as anxious as Gregory Bennett to get the real transplant. The Frintons and the Bennetts got together and decided to take legal action. They hired Charles Frinton's former law firm to write letters presenting the legal case for Frinton and Bennett's right to receive immediate transplants, and threatening further action if they did not. The letters went not just to Fahn and myself, but also to the NIH, the board of Columbia-Presbyterian Medical Center, and the University of Colorado. Since the law firm billed at nearly $300 per hour, the legal effort cost several thousand dollars.

Besides these legal efforts, the Bennetts and Frintons also wrote personally to their elected representatives, and Pauline made repeated phone calls to the NIH. Eventually, around April 1998, the NIH agreed that the sham-operated patients were legally entitled to makeup surgery without waiting for the completion of the study, even if there was a safety concern. The NIH therefore overrode the PSMB once again. Even then, Gregory Bennett had to wait. First, there were two other patients ahead of him in line, including Charles Frinton. And then there were further delays because of difficulty in finding properly matched tissue. He finally had his real surgery in October 1998.

The operation went uneventfully, although Bennett did have a puffy face for several days after surgery. This soon corrected itself.

After that, nothing much happened for several weeks, but in January 1999 Bennett noticed that his tremor was diminished and the frequency of on-off fluctuations had lessened. The improvement continued, and Bennett was eventually able to reduce his L-dopa dosage by half. This led to a marked lessening of his dyskinesias. His gait and his handwriting also improved but did not return completely to normal.

By the time Sid Howard approached his one-year anniversary, he and his wife, Nee, had gone from thinking he had had sham surgery— their opinion right after surgery—to thinking he had had a real transplant. In spite of the setback caused by his heart attack, they felt that his neurological condition was distinctly improved in some respects, and the objective tests at one year confirmed this. They were expecting Paul Greene to tell them that Sid had had a real transplant, and this is in fact what he did tell them. As with Emily Mason, however, there was a long delay before they found out. By the time they got the news, Sid's anniversary was six months behind him.

After that time, Howard continued to do reasonably well. The most striking benefit has been a marked reduction in his dyskinesias. "I see them occasionally, when he's forgotten to take his medication and he's tired," says Nee. "Before, he had them wherever he went—they would attract attention and were very embarrassing for him."

Jack Celnik's anniversary was in January 1998, but it took another four months before he got the official word on whether he had received a real or sham transplant. When he called to find out the reason for the delay—which he did repeatedly—he was told that there were problems with the transfer of documents from the North Shore Hospital, where his one-year PET scan was done, to Columbia. Then he was told that key personnel who needed to analyze or sign off on the data were away—the statistician, for example.

Celnik was so convinced that his operation had been a sham that he didn't wait for the official word. Well before his one-year anniversary he learned about the PSMB's decision to delay the makeup surgeries and became involved in the effort to overturn it. He was recruited into the effort by other patients in the study. Celnik's role, however, was fairly minor: he made contact with some members of Columbia University's board of trustees, explaining the patients' point of view to them.

As the months rolled by without the official word, Celnik became more and more frustrated. But more problems lay ahead. In May 1998 Greene called Celnik and confirmed what he believed: yes, the operation had been a sham. But he also told Celnik that, although he had a right to a real transplant, he should wait another six months before making a decision. That was because we were preparing to undertake a preliminary analysis of the results to date, based on the patients who were already unblinded, and we hoped to be able to let Celnik and the other sham-operated patients know what we found. That way, even if they didn't want to wait for the full analysis of the data, they could still make a more informed decision than they could without any notion of the study's outcome.

Celnik did agree to wait, but when we had the preliminary results, they didn't bode well for him. It was beginning to look as if the benefits of the fetal-cell transplants were confined to the younger group of patients—those under sixty. Celnik had been just under sixty when he first applied to the study, but now, in 1999, he was sixty-three.

Of course, there's nothing especially fateful about someone's six-tieth birthday. Our preliminary results didn't mean that a person who turns sixty changes overnight from a "responder" to a "non-responder" to fetal-cell therapy—far from it. There was consider-able variability among individuals, even among those of the same age. It just so happened that we had used the age of sixty to divide our patients into two groups, and Celnik now fell in the older group that didn't seem to benefit from the transplants.

Celnik talked with Greene and Fahn and myself about his options, and unfortunately he got rather mixed messages. Greene and Fahn told him in fairly strong terms that he should not opt for a transplant, because the chances of a detrimental outcome out-weighed the likely benefits. They mentioned a new treatment that involved the implantation of an electrical stimulator in his brain. But, they said, this new treatment was still experimental, and Celnik should hold out until it was better proven.

When Celnik talked with me, on the other hand, I suggested that he might still want to opt for a transplant, given his young age within the older group. Also, I had not given up on the possibility that the transplants would eventually turn out to be beneficial in the older patients. On a second occasion when I talked with him I mentioned that Bob Breeze and I were working on a new version of the trans-plant procedure that we felt might offer greater benefit and less risk of harm than the one we had been using in the double-blind study. (I'll have more to say about this in chapter 20.) So again, I suggested that he wait until that procedure had been developed.

Celnik, whose disease had progressed noticeably during the time he had been involved in the study, now felt that his chances of getting any kind of relief were rapidly disappearing. He wrote a letter to Fahn and myself that took the form of a parable about a man crawling across the desert in the last extremity of thirst. An angel, he wrote, held out a glass of water, but every time the man tried to approach it, the angel moved away and told him to crawl farther. Finally, both the angel and the water turned out to be a mirage.

The letter was an eloquent expression of Celnik's emotional state at the time, and it was impossible not to sympathize with his sense of disappointment. A part of me wanted to tell him: Come and get your transplant surgery tomorrow! But the fact was, the limited information available at the time did not support the idea that he would benefit from the surgery. I could not in good conscience urge him to have the transplant; I could only present it to him as an option to which he was entitled. Celnik decided that he should wait, and he has not had transplant surgery or any other procedure since that time.

James Ross had vacillated in his opinion as to whether his operation had been a real transplant or a sham: the operation itself had seemed real, and he and Gail sensed some improvements over the subsequent few months, but the improvements were not such as to greatly enhance the quality of his life. When he came to the time of the official disclosure in the fall of 1998, he wasn't sure what to expect.

Paul Greene told him that he had had a real operation. To some degree, the news disappointed Ross, because the benefits of the surgery had been much less than he had hoped. Still, he remains pleased by the modest improvements he has experienced. "I'm surprised how good I am in the part of the tests where I'm completely off drugs," he says. "I don't have any tremor, and I can do things still. I thought I'd be completely laid out." Still, when Ross is on medications he experiences dyskinesias that can be quite severe at times. He judges his physical condition by his ability to get a basketball through the hoop in the forecourt of his house—sometime he can and sometime he can't.

Jacqueline Winterkorn's condition improved so much during the year after her operation that she was convinced she had a real transplant well before the time of her official notification. When the time came around, Paul Greene asked her whether she would mind if a TV crew filmed her receiving the news, for a documentary film they were making on the transplant surgery. She agreed, and the crew came to her office and had the cameras running when she got the call. But it wasn't all that dramatic, because Greene told her that her operation had indeed been a real transplant, and Winterkorn was completely unsurprised.

Since that time, she has improved even further. "My gait is wonderful," she says. "I haven't had any freezing—knock on wood—I can get up in the middle of the night and go to the bathroom or walk to get ice cream. I never have to worry about shuffling or hurried gait—I have a good stride." Her handwriting has also improved markedly. "For five or so years before my transplant I hadn't been able to send Christmas cards; my husband had to write them all. Last year I sent out 100 cards and wrote notes on half of them. I can write thank-you notes and take messages from the cell phone and so on. I'm still annoyed because I used to have gorgeous handwriting and now it's sloppy, but it's perfectly legible."

By one year after the transplant, Winterkorn had reduced her L-dopa dosage from 1.2 grams per day to 0.8 grams per day. Since then, she has gone down to 0.4 grams per day—one-third of her preoperative level. This dose reduction has been accompanied by a marked reduction in her dyskinesias, though they have not disappeared completely. "I get the 'heeby-jeeby' type, where I shift from foot to foot and move around as if I'm really antsy," she says. "If I went back on 1.2 grams of L-dopa a day I would be dyskinetic off the wall." The only deterioration in her condition that she notes is a problem with her speech while she is "off"—the labor of moving her mouth makes it difficult for her to enunciate well. When she is "on," on the other hand, her voice is as loud and clear as one could wish—a far cry from the mangled speech that so interfered with her work prior to her transplant.

CHAPTER 19

Decoding

As soon as Jacqueline Winterkorn was "unblinded," in late January 1999, we could proceed with the long-awaited analysis of the overall data. The lead statistician, Wei-Yann Tsai (who had taken over from Bill DuMouchel during the course of the study), had already started the number-crunching before Winterkorn's disclosure date. Within a couple of weeks he was able to fax us a mass of data, and in early February we had a teleconference to review the results. Fahn, Greene, Eidelberg, Tsai, Dillon, Winfield, and some others were in a conference room at Columbia, and Breeze, myself, and another statistician, Jim Murphy, were in Denver. Over the course of three hours we looked over the graphs, histograms, and statistical analyses that Tsai had prepared and discussed what they meant.

We already had some notion of what the results would be. The New York group knew that some of the patients who had received transplants, such as Jacqueline Winterkorn, had done well, while others, such as James Ross, had not. It was also clear that the success stories were concentrated among the younger patients. The statistical analysis was another matter altogether. This was really the central point of the study: could we document a benefit of the transplant procedure in a completely objective way, looking at entire groups of patients rather than just our "star performers"?

Somewhat against my better judgment we had defined the Global Rating Scale—the patients' self-assessment of how improved they were one year after surgery—as the "primary outcome variable," the principal measure of the efficacy of the transplant procedure. In the older group of patients—those over sixty—there was no difference between the GRS scores for the experimental and the sham-operated patients: both showed a small improvement. In the younger group of patients, the GRS scores were higher for the experimental group than the sham-operated group: the experimental group had an average score of about 1 ("slightly improved"), whereas the sham operated group had an average score of about 0.3. But given the small numbers of patients, this difference was not large enough to achieve statistical significance. Thus, our principal measure failed to show a significant benefit of the transplant procedure in either the older or the younger group of patients.

This was disappointing, certainly, but it wasn't totally unexpected. I had long known that only a minority of transplant recipients—perhaps one-third—do so well that they would be likely to check the "greatly improved" box. Some recipients do not improve at all and some deteriorate. Thus, an average score in the neighborhood of 1 was to be expected.

Naturally, the negative result with the Global Rating Scale reinforced my sense that it was not the optimal measure to judge the efficacy of the transplant. I became that much more interested in the results of the objective tests, which directly assessed the movement problems that are at the core of Parkinson's disease. Here the UPDRS was the key measure.

You may recall that Sandra Dillon and Hal Winfield tested the motor component of the UPDRS in two conditions: when the patients had gone overnight without drugs ("off"), and after their usual dosage of L-dopa ("best-on"). The results of the study showed a significant improvement in the "off" scores in the younger group of patients who received transplants, compared with the younger sham-operated patients. Thus, as a group, younger Parkinson's disease patients do benefit from fetal-cell transplants in the core aspect of their disease—their ability to move. Looking at the individual symptoms, we found that the transplants were beneficial for two of the three cardinal signs of Parkinson's disease, rigidity and slowness (bradykinesia), but not for the third, tremor.

For the "on" UPDRS scores, there was no significant difference between the real-and sham-operated patients in the younger age group. In other words, when the patients' L-dopa was working optimally, the transplants did not confer any significant additional benefit.

Another important measure that showed a benefit of the transplant surgery in the younger group was the Schwab and England "Activities of Daily Living." By the end of the first year, the transplant recipients were scoring significantly better than the sham-operated patients in the Schwab and England "off" scores, meaning that they were able to carry out day-to-day tasks in the "off" state better than were the sham-operated patients.

The story with the older patients was not so positive. There was no significant difference between the real- and sham-operated older patients in any of the measures at one year after surgery, although on several of the measures, such as the UPDRS "off" scores, there was a tendency for the transplant recipients to score better.

Why were the results less good in the older than in the younger patients? We know from the PET scans and the autopsy results that the grafts do survive and grow in the older patients, just as they do in the younger ones. It may be, however, that there is insufficient plasticity in the brains of the older patients to take advantage of that new growth. Another factor may be that the Parkinson's disease in older patients is somewhat different from the disease in younger people: in particular, it may affect a number of brain systems rather than being limited to the striatum. One reason for suspecting that this is the case is that Parkinson's disease is more likely to be accompanied by cognitive or memory problems in older patients, suggesting that the cerebral cortex may also be involved.

To the extent that other regions besides the striatum may be involved, the addition of new cells to the striatum may be a less effective remedy.

Yet another possibility, however, is that one year was simply not long enough to see beneficial effects of the transplants in the older patients. It may be that the brains of older persons adapt more slowly to the presence of the transplanted fetal cells than do those of younger persons. This in turn made me wonder whether, over longer periods of time than a single year, significant benefits might begin to emerge. That did indeed turn out to be the case, which I'll get to later.

So, the results at one year showed that the fetal-cell transplant did have beneficial effects in the younger patients, although the effects were not so striking as to lead to a clear difference in the patients' self-assessed state as compared with the sham-operated patients. Were there any negative effects of the transplants? I've already mentioned the matter of the "serious adverse events." In the complete one-year analysis, eight such events occurred in the experimental group and only one (Emily Mason's hysterectomy) occurred in the sham-operated group. Did this mean that the transplants caused the excess of adverse events? We don't know for sure, but I think not. Except for one event (the confusional episode caused by a subdural hematoma) there simply were not any persuasive causal connections between the transplant surgery and the particular kinds of adverse events that the patients suffered. Most likely, the imbalance in the distribution of the adverse events was simply a matter of chance.

There was one other potential ill effect of the transplants. Of the 34 patients who eventually received transplants in the study (the original 20 in the real-transplant group, plus 14 of the sham-operated group who later requested and received makeup surgery), five developed quite severe dyskinesias at some point after the transplant, even after reducing or stopping their L-dopa. Three patients were so severely disabled by the dyskinesias that we had to offer them a different kind of surgery (the placement of an electrical stimulator in the globus pallidus) in an attempt to reduce the dyskinesias. The stimulating device was implanted in the summer of 2000.

Was the development of severe dyskinesias in these five patients simply a matter of chance—dyskinesias are common in Parkinson's disease patients, after all—or did they develop as a consequence of the transplant surgery? These patients had had L-dopa–induced dyskinesias prior to the transplant. Dyskinesias generally are a side effect of L-dopa treatment—they probably result from the chronic, unregulated exposure to dopamine that L-dopa therapy provides. The fetal-cell grafts also release dopamine, and how well regulated that release may be is not yet known. Stanley Fahn and Paul Greene therefore take the view that the five cases with severe postoperative dyskinesias are a negative consequence of the fetal-cell transplants, and should be weighed in the balance against the beneficial effects already mentioned. Because the transplants are releasing dopamine, they are capable of producing the same side effects as L-dopa. I believe that the risk of dyskinesias can be reduced by changing the placement of the grafts in the brain. It's worth bearing in mind that many of the transplant patients experienced a *lessening* of their pre-existing dyskinesias.

We would have plenty of opportunity to discuss all these issues over the ensuing months, but the first order of business after going through our results was to communicate them to other people. The people who needed to know first were the remaining sham-operated patients who had not yet received makeup surgery—Jack Celnik among them. We told them the results but asked them to keep the data confidential until we had made a public announcement. The results documented what we had already hinted to Celnik, namely that a transplant was not likely to help him, given his present age. Thus, he did not elect to get makeup surgery, and neither did the other patients who were waiting, even the two patients who were in the younger group.

To some extent, this proved the sense of the point of view taken by the Performance and Safety Monitoring Board, when they had argued that the sham-operated patients should wait until the end of the study before deciding on makeup surgery. However, the fact remains that some of the patients who got makeup surgery, such as Emily Mason, did well, and the entire group of younger patients who got makeup surgery showed the same average improvement, in terms of UPDRS and Activities of Daily Living scores, as did the younger group of the original transplant recipients. So if their insistence on getting the makeup surgery was a gamble, it was a gamble that turned out to be successful for many of them.

We planned to present our results at the annual meeting of the American Academy of Neurology, which was to take place in late April in Toronto, and much of the intervening time was spent getting the data into a form suitable for presentation. Mike Walker, the NIH administrator who oversaw our grant, wanted to see the data before it was announced, so we scheduled a meeting in New York for April 8. Before that meeting could happen, however, I had a "serious adverse event" of my own.

On the afternoon of April 2, which was Good Friday, I set out to drive from Denver to Vail. My wife and I both enjoy skiing and we often spend weekends in Vail, where we have a condominium. I had barely reached the foothills of the mountains, however, when the pickup truck I was driving hit an icy spot on the freeway, skidded sideways across three lanes of traffic, went down into a ditch, and turned over. The roof of the pickup was crushed right down to the level of the steering wheel. My head was driven through the window, and I was knocked unconscious.

When I came to, paramedics were attending to the top of my scalp, which had been sliced open almost from ear to ear and was bleeding profusely. I was so confused I didn't know where I was, and I couldn't understand how my ski gear came to be scattered across the road. As the paramedics put me on a board and lifted me into the ambulance, I called out, "Where's my wife? Where's my wife?" That sent two of the rescue workers rushing to see if she was still pinned under the vehicle. It was only when the ambulance was halfway to Denver, and my memory was starting to clear, that I recalled that my wife had not been with me. She was recovering from knee surgery, so skiing was not an option for her on that particular weekend—luckily.

Although I lost about a third of my blood and required a long row of staples to close the scalp wound, I was able to leave hospital the following day. In fact, I left for New York on the following Tuesday as scheduled. We were to go over the clinical data with Fahn's group and representatives from NIH who wanted to know what we had found. Coincidentally, ABC's *20/20* was doing a story on placebo surgery and asked to interview Paul Greene and me. It was a bit comical to be talking about brain-cell transplantation while sporting a Frankenstein-like row of staples across my own head. Luckily, we were able to find a camera angle where the staples weren't too evident.

On the following day we all met with Mike Walker and gave him a complete rundown on the results of the study. Because of the importance of the study, as well as its political overtones, Walker decided to come to the Toronto meeting to represent the NIH's point of view. Then it was back to Denver to get ready for the meeting.

The meeting itself was one of the high points of my career to date. My colleagues and I had the opportunity to present the results of the first-ever scientifically controlled study of a therapy aimed at repairing the damaged brain. There was huge interest among the neurologists gathered for the meeting, an interest that had been stoked in part by a news article in the *New York Times* the day before our talks. Before our presentations I had to constantly dodge questions from colleagues and from media representatives about what our results had been.

We had been allotted one ten-minute oral presentation as well as three poster presentations. The oral presentation, however, was slated for the end of a morning session, which seemed like tacit permission for us to go on longer than the usual ten minutes—since we wouldn't be holding up anyone else's talk. So Stan Fahn and I split the presentation: I talked for ten minutes about the background and design of the study, the technicalities of the transplant procedure, and the autopsy results, then Fahn talked for another ten minutes about the results. The moderator, who has the responsibility for hauling speakers off the stage when they exceed their allotted time, let us speak for a total of about twenty minutes.

The talks were very well received and immediately afterward we held a press conference to make the results available to the media. Two of our transplant recipients who lived near Toronto, Emily Mason and another woman named Lynda McKenzie, attended the press conference. They didn't speak, but we mentioned their presence and that they were willing to give interviews afterward. The press did indeed descend on them. I believe they were interviewed for hours, and stories about them appeared widely over the following few days. While Lynda had only recently received a transplant and had no change in her condition, both women expressed themselves positively about the study. Although these were the only two patients at the Toronto meeting, several other patients made themselves available for interviews in that same time period, including some who did not benefit or who got sham transplants and then declined makeup surgery. Jack Celnik, for example, was interviewed by the *New York Times*, and his disappointing personal experience was evident in the resulting article.

The following day, interest in the study continued unabated. We presented three posters, one concerned with the overall study data, one with the PET results, and a third devoted to the longer-term data—that is, the data on the patients who got their transplants early in the study and were now two years or more post-transplant. All of us involved in the study got a lot of congratulations, not just on the positive results of the study but even more on the effort and care we had put into making it truly objective and scientific.

The completion of the blinded phase of the study and the presentation of the one-year results did not mean that the study was over—far from it. We planned to continue studying all the patients for at least another two years. Would the patients who had benefited from their transplants continue to do well, or would the benefits wear off? The latter was at least a theoretical possibility, for two reasons. First, the progression of the patients' underlying disease, and the resulting continued death of dopamine cells, might put a greater and greater burden on the grafted cells—a burden that they might eventually be unable to meet. Also, whatever process caused the death of the patients' own dopamine cells might eventually attack and kill the grafted cells, too. Finally, there was the possibility that the grafted cells might eventually be rejected by the patients' immune systems.

We also wanted to know whether the patients who did *not* show significant benefits from their transplants would do so at longer times—perhaps after two, three, or more years. We thought that this was a particularly likely possibility for the older patients, because even in the one-year data there was a suggestion of a benefit, although it didn't reach statistical significance.

Finally, we wanted to know if any harmful consequences of the transplants would show themselves. I already mentioned the possibility that the transplants had caused the appearance of severe dyskinesias in a few patients. If so, it was conceivable that more and more of the transplant recipients might develop dyskinesias as time went by. We needed to keep track of these and any other potential ill effects.

My first order of business after the April 1999 meeting in Toronto was to write an application for a two-year renewal of the NIH grant. This took me until the beginning of July. The grant renewal was funded, so the study is still continuing today, and our loyal patients are still making their trips—now twice a year rather than three times—for several days of exams at Columbia-Presbyterian Hospital. How many more times are we going to have to tap the floor with our heels, they're probably thinking. How many more times do we have to count down from 100 by sevens? Most of the patients can probably recite the correct sequence of numbers in their sleep at this point. But these patients are a uniquely valuable resource—only they can tell us what the long-term outcome of fetal-cell transplantation in Parkinson's disease may be.

The grant application done, I turned my attention to another major task—writing up our results so far for publication. We decided to send an initial manuscript to the *New England Journal of Medicine* that gave a general description of the study and its results up to one-year post-transplant, and to follow that up later with papers that would cover specialized aspects of the study, such as the PET results. Those papers would go to specialty journals.

Writing the *NEJM* paper turned out to be a Herculean task. I began in July of 1999. Stan Fahn and I swapped versions repeatedly over the following few months, trying to reach agreement on how the data were to be presented. To some extent, the issues were simply technical or stylistic, but Fahn and I also differed somewhat in our interpretation of the study's results, with Fahn taking a more negative view than I do. The matter of the postoperative dyskinesias was an example of that. But more generally, Fahn thought that the results were not good enough to encourage the introduction of fetal-cell transplantation as a conventional treatment for Parkinson's disease, and that the paper should therefore not be seen as making a pitch for such a treatment.

Actually, I was in complete agreement with Fahn on that point. But as someone who has devoted more than fifteen years to the hypothesis that we can repair damaged brains, I saw the study as a crucial stepping-stone toward finding the way to that goal, not as the goal in itself. Thus I was more inclined to stress the study's positive results, rather than its limitations. It was, after all, the first double-blind neurosurgical study ever conducted, and it showed highly significant improvement in the key symptoms of Parkinson's disease, at least in the younger patients. From the point of view of a clinical trial of only forty patients, ours was a remarkable success.

Because of these differences, the manuscript went back and forth between us quite a few times. It wasn't until December that he and I (and the other authors, who included Bob Breeze, David Eidelberg, Paul Greene, and quite a few others) reached a consensus and were able to send the manuscript off to the *NEJM*. Then the journal sent the manuscript back with a number of requests for changes. In the meantime, Mike Walker of the NIH got involved: he wanted us to include more data about the long-term follow-ups. So we had to make several more revisions.

The paper finally appeared in the *New England Journal of Medicine* on March 8, 2001, amid another storm of contradictory publicity. The same currents of politics, egos, and science that we experienced in 1988 have not lost their force. We proved many things. We found that transplants grew in 85 percent of patients without immunosuppression regardless of age. Another clear result of our study was that only the patients under sixty showed significant improvement as a group, even though about one-third of the older patients also improved to some degree. An unexpected result was that the operation seemed likely to benefit men more than women.

Several questions were answered, but many more have been posed by our study. I truly thought we would have perfected transplants by the year 2000. We've made progress, but a predictable treatment remains elusive.

CHAPTER 20

The Future

The methods we used to carry out the fetal-cell transplants in the double-blind study represented the state of the art in 1994. After that, we remained locked in to that technology for the duration of the study. Changing anything would have put the scientific value of the study at risk and would also have required us to submit a revised application to the NIH and undergo further lengthy reviews.

After we completed all the transplant operations and makeup surgeries, we were free to head off in new directions. There were indeed reasons to consider changes in the transplant technology. In part these emerged from the double-blind study, and in part they derived from laboratory research, done by ourselves and others, that hinted at new therapeutic avenues which needed to be explored.

The double-blind study confirmed what I had long known: some patients do spectacularly well after receiving dopamine-cell transplants, some show modest improvement, and some do not improve at all. Obviously, we would like to be able to tell in advance which patients are likely to benefit. That way we could improve the prospects for those patients who received transplants, and we could spare other patients the stress, risks, and expense of an operation that was not going to help them.

What does decide whether a patient will benefit from a transplant? The double-blind study identified one factor: age. Still, the age factor was a bit puzzling. Some people up to age seventy-five did very well, while others did not. We've taken a closer look at what is going on and have come up with a striking observation. What we've found is that age is *not* the important factor. Instead, it was the patients' response to L-dopa *before* the surgery that best predicted whether they would have a good response to the transplant.

Everyone who participated in the study had to have some response to L-dopa to qualify. The minimum response was a 30 percent change in the standard UPDRS test from the "off" state first thing in the morning to the "on" state. All of the patients in the younger group had a 60 to 95 percent response to L-dopa (a 95 percent response meant essentially complete disappearance of Parkinson symptoms) while the people older than sixty were more variable, some with only 30 percent, others up to 90 percent. We found that the older patients with an excellent response to L-dopa improved just as much as the younger patients, while those with a minimal response to L-dopa were not improved by the transplant. It did, however, take longer for older patients to show a significant response, two years instead of one.

The reason some people do not respond to L-dopa is probably because they have lost more than just the dopamine neurons in the brain. Other areas beyond the striatum are probably involved, so solving the dopamine shortage in the striatum is not sufficient to provide a clinical benefit.

Even for younger patients with good responses to L-dopa, not everyone improved. So what other factors are at play? There are many possibilities. One likely factor is the kind of Parkinson's disease that affects the individual patient. Some patients (like Emily Mason, for example) have a family history of the disease and thus may have inherited genes predisposing them to the condition; others have no such history and may have developed the disease on account of some environmental factor or for other still unknown reasons.

Another crucial factor affecting the success of the transplant has to do with the tissue that is transplanted. We know the developmental age of the embryos being transplanted, and we also know how much dopamine the cells are making. The biggest uncertainty is how many of the cells will survive after transplantation. We know that at least 90 percent of the dopamine cells do not survive. This has been true in all of the animal research and in humans who have been autopsied. Still, from the PET scans and the autopsy results, significant numbers of grafted cells do survive and grow in the great majority of patients, even in those who do not show clinical benefit. Experiments we have done in rats have shown that treatment of the dopamine neurons with chemical growth factors *before* the transplant can double the number of cells that survive. We are seeking approval from the FDA to use the growth factors in tissue culture before transplant, but their regulatory hurdle is very high.

We are also adding to the brain locations where we make our transplants. Up until recently, we have made all our transplants into the striatum. More precisely, we have made two needle passes into the division of the striatum known as the putamen on each side of the brain. That's because the putamen is more severely affected by dopamine loss in Parkinson's disease than is the other division, the caudate nucleus.

Recall, though, that the normal location of dopamine cells is not the striatum at all, but the substantia nigra, which is located in the midbrain. In a healthy person, dopamine is supplied to the striatum not by cell bodies located in the striatum itself, but by the nerve endings of the long axons of the nigrostriatal pathway. The dopamine cell bodies in the substantia nigra not only send fibers to the striatum, they also distribute a network of fibers and terminals within the substantia nigra itself. In Parkinson's disease this dopamine network within the substantia nigra degenerates, just as the one in the striatum does. But transplanting dopamine cells into the striatum of a Parkinson's disease patient does nothing to replace the network within the substantia nigra. Would it help patients to replace that network, too?

Several groups of researchers have studied the effects of transplanting dopamine neurons into the substantia nigra of laboratory animals, especially in rats whose nigrostriatal pathways have previously been destroyed by injections of the neurotoxin 6-hydroxydopamine (6-OHDA—see chapter 4). In general, a transplant into the substantia nigra does not provide nearly as effective relief of symptoms as does transplantation into the striatum. The main reason for this is that the nerve fibers put out by the transplanted cells don't form a new nigrostriatal pathway—they stay within the substantia nigra.

More exciting results have been obtained in a recent series of studies done by neurosurgeon Ivar Mendez and his colleagues at Dalhousie University in Nova Scotia. Mendez's group transplanted dopamine cells into *both* the striatum and the substantia nigra of 6-OHDA-treated rats. They found that the benefits of these double transplants were greater than those of transplants into either site alone. Not only did the rats' circling behavior correct itself faster and more completely, they also improved in some other symptoms that are not alleviated by transplants into a single location.

But the most interesting observations came when Mendez's group sacrificed the rats and examined their brains. As they expected, the dopamine cells at both graft sites had survived and put out fibers locally. Unexpectedly, however, some of the fibers put out by the dopamine cells in the substantia nigra had grown all the way to the striatum and formed a network of terminals there. Somehow, the presence of dopamine cells in the striatum had encouraged the transplanted cells in the substantia nigra to do what they would not otherwise have done, namely to regenerate a nigrostriatal pathway.

These results do not mean that similar double transplants in Parkinson's disease patients would necessarily result in the formation of new nigrostriatal pathways. The distances involved are much greater in humans than in rats. Even so, the results from the

Dalhousie lab and from other researchers suggest that such double transplants may offer extra benefits over transplants limited to the striatum, even if the nigro-striatal pathway does not re-form.

So far, we have tested this new strategy in one human patient with Parkinson's disease. After getting permission from our Institutional Review Board, Bob Breeze and I performed the operation in October 2000. In the new procedure, we made only a single pass into the putamen on each side instead of the two passes that we made in our earlier patients. The passes were positioned in such a way as to avoid the region of the putamen that we believe is most liable to trigger dyskinesias after transplants. Then we made a single pass into the substantia nigra on each side, placing dopamine cells there too. By one year after the operation, the patient had improved enough that he was able to resume his favorite pastime, golf, and was shooting rounds in the low eighties—as good a score as he had ever achieved before his illness. His PET scan shows that the dopamine cells are surviving and developing at all the transplantation sites.

My guess is that the double transplants will provide some extra benefits in humans, but I doubt that they will lead to adequate regeneration of the nigrostriatal pathway or to a complete cure for Parkinson's disease. If this turns out to be the case, we will be driven to the conclusion that the anatomical integrity of the nigrostriatal pathway is vital for normal functioning. And such a conclusion would hardly be surprising: after all, dopamine cells transplanted to the striatum lack some of their normal input connections, namely the inputs they would receive if they were in their normal location in the substantia nigra. Without these inputs, the transplanted cells can't be expected to release dopamine in a completely normal fashion. If all this turns out to be the case, we will have to turn our attention to the difficult task of rebuilding the normal nigrostriatal pathway.

Neuroscientists have been struggling for decades to induce the brain to regenerate its long fiber pathways. Achieving such a goal is the Holy Grail of neurology; it would offer incalculable benefits to the victims of spinal cord injuries, stroke, and many other devastating neurological conditions. But as yet no one has achieved a substantial functional reconnection of a long fiber system.

In the case of the nigrostriatal pathway, there have been hints of success. I already mentioned the Dalhousie group's observations on the doubly transplanted rats. Another group, led by Feng Zhou of the University of Indiana, got promising results with another strategy. They reasoned that dopamine cells grafted into the substantia nigra don't grow fibers toward the striatum, not because of any deficiency in the cells themselves, but because the pathway along which the fibers would have to travel is unfriendly to nerve growth. Either it is physically difficult to penetrate or (more probably) it lacks the chemical signals that encourage growth during fetal life, when the brain is first assembling itself.

Zhou and his colleagues tried the following experiment. First they gave rats the standard 6-OHDA injections into the substantia nigra, killing the dopamine cells and creating the movement disorder. Then they constructed an easy-to-navigate pathway between the substantia nigra and the nigra by a radical strategy: they injected tiny doses of a toxic chemical along the pathway. The chemical killed nerve cells in the vicinity of the injections, forming a "trail of destruction" between the substantia nigra and the striatum. Finally they transplanted fetal dopamine cells into the 6-OHDA treated nigra. The transplanted cells did indeed send out fibers along the trail of destruction, and the fibers did partially reinnervate the striatum. This in turn permitted a partial restoration of function.

Why did the "trail of destruction" promote fiber growth? It's known that damaged brain tissue releases a variety of hormonelike molecules called "growth factors" that encourage the development and survival of nerve cells and their connections. Several of these growth factors have been identified and in fact can be purchased from specialty pharmaceutical companies. So could one achieve the same effect, not by destroying cells, but simply by injecting growth factors? Such an experiment was tried by none other than Barry Hoffer—the Denver neuroscientist who protested rather noisily against the transplant we performed on Don Nelson in 1988. Hoffer (who is now director of intramural research at the National Institute of Drug Abuse), along with several colleagues, did an experiment almost identical to Zhou's. Instead of creating a trail of destruction, however, they simply made a row of injections of a growth factor along the pathway. The particular growth factor they used is called "glial cell-line derived neurotropic factor" or GDNF, and it is a molecule that is normally synthesized by the non-neuronal supporting cells of the brain, known as "glial cells." Hoffer's group got a result similar to Zhou's: the transplanted dopamine cells sent out fibers along the GDNF-treated pathway and partially reinnervated the striatum. Obviously, a therapy involving injections of a growth factor comes closer to a possible clinical therapy than one involving injections of a toxic chemical.

I describe these experiments not to tout them as instant therapies for Parkinson's disease or any other disorder, but as examples of the many promising avenues that are currently being explored. At least one of them will eventually lead to the desired goal, I'm convinced, and that will be very good news for millions of people.

There's another whole line of research that has a somewhat different goal, which is to find a substitute for the cells we now use for transplantation. In themselves, dopamine cells derived from human fetuses are probably close to ideal, but the source of the cells—elective abortions—creates numerous problems. First, it presents a moral quandary for some patients. Second, the abortion procedure is designed to destroy the fetus and leaves it in small fragments; only in a minority of cases do these fragments yield usable tissue. Since we need tissue from several fetuses for a single transplant, we often have to wait weeks to get enough. Third, there is an ever-present risk of infection; although it hasn't happened yet, we are always concerned that virus- or bacteria-contaminated tissue could slip past our screening procedures and cause a serious infection in the recipient. And fourth, there is the problem of tissue-matching the fetal cells to the recipient. Thus, even though upwards of a million abortions are performed in the United States every year, it's not realistic to think that one could ever obtain enough tissue to perform transplant surgery on all the patients who might benefit from it.

What other kinds of cells might offer an alternative? One possibility that has already reached the stage of clinical testing is the use of dopamine cells taken from animals. Experiments by Ole Isacson at Harvard had shown that fetal pig dopamine cells could survive and improve signs of Parkinson's in animal models. Based on those observations, the FDA gave the okay for an initial study of twelve patients to see if the foreign tissue grafting was a reasonably safe procedure. After that, a larger, double-blind study with a design similar to ours was done. The cells were prepared by the Diacrin Corporation in Boston, and the transplant surgery was done at the University of South Florida and at Emory University in Atlanta. Patients with advanced Parkinson's disease were enrolled; half received transplants of the pig cells and the others had sham operations. Because rejection of the pig cells would have been certain without immunosuppression, all patients received the drug cyclosporine.

When the blind was broken in March 2001, there turned out to be no benefit to the pig cell transplants. The PET scans showed that the pig cells had not survived. The patients' immunologic attack on pig tissue wiped out the cells, overwhelming the effect of the immunosuppressant drugs. The immunologic barrier of foreign-species transplants remains a daunting challenge. Even if drugs could be shown to prevent rejection, the necessity for lifelong immunosuppression has to be counted as a significant drawback to the procedure.

Another theoretical risk is that certain retroviruses found in pigs might cause health problems in the transplant recipients. So far nothing like that has happened. Still, our experience with the AIDS epidemic, which is caused by a retrovirus that seems to have jumped from chimpanzees to humans, should alert us to the possibility of something similar happening between pigs and humans.

Because human fetal tissue is difficult to obtain and animal tissue is unlikely to survive transplantation, I am pursuing a number of other strategies. One idea we are investigating in Denver is to take fetal dopamine cells and make them divide repeatedly in tissue culture, thus providing a limitless supply of cells for transplantation. Normally, the fetal dopamine cells that we take from the seven-week fetuses do not divide—neither in tissue culture nor in the patient's brain after we transplant them. However, my Denver colleague Kedar Prasad has been able to make the cells do so by genetic trickery—specifically, by inserting a viral gene into them. The altered cells divide about once a day, meaning that under optimal conditions a single cell could be turned into a million cells in a mere three weeks. If Kedar's or another lab finds the secret for producing the authentic brain dopamine neuron, a few months' production could create enough cells to provide transplants for every Parkinson's disease patient in the world.

Of course, it isn't that simple. One major concern is that we don't want the cells to go on dividing once they have been transplanted in a patient—otherwise they would constitute a brain tumor and might well kill the patient. What we have found so far in rats, however, is that the cells do not divide after they have been transplanted. Somehow the brain "tells" the transplanted cells that they should stop dividing. Interestingly, when we biopsy the grafted cells, taking them out of the host rat and putting them back in tissue culture, they start dividing again.

Another concern is whether Prasad's "immortalized" cells will continue to display all the properties of dopamine cells when they are transplanted. Our studies to date suggest that they retain some properties but lose others. In rats injected with 6-OHDA to give them a movement disorder, transplants of Prasad's cells relieve some of the symptoms. Whether the transplanted cells will be sufficiently like normal dopamine cells to replace the functions of those destroyed by Parkinson's disease is a question that can be answered only by further research.

The whole field of transplantation is likely to be revolutionized by exciting new work on cells called *stem cells*. There are many kinds of stem cells. Some are present in most of our body organs throughout life. Every day, stem cells in our bone marrow must produce huge numbers of white blood cells, red blood cells, and the clotting cells called platelets. Similarly, the skin as well as the lining of the intestine are continuously replaced with new cells arising from stem cells in the skin and intestine.

Recently, politicians have given these kinds of stem cells the name "adult stem cells" to distinguish them from "embryonic stem cells." Embryonic stem cells are cells that are generated from embryos produced in fertility clinics. These embryos, which consist of a few hundred cells, have usually been frozen for months to years. Eventually, the couple who generated the embryos to have their own children must decide what to do with the surplus frozen embryos they no longer want. Most often, the embryos are thawed and discarded. Sometimes, the couple may decide to donate the frozen embryos for medical research.

Those donated embryos are the source of embryonic stem cells. The cells can be put into tissue culture and can divide indefinitely, forming billions of identical cells. What is remarkable about embryonic stem cells is that they can become any tissue in the body and are not limited to becoming skin, muscle, or brain cells.

The problem with stem cells, whether from adults or from embryos, is that we have only begun the work needed to understand how to convert them into specific kinds of cells needed to treat disease. For treating Parkinson's disease, for example, dopamine cells must be developed. For diabetes, we need cells that can produce insulin when needed.

The cells that give rise to brain tissue are *neural stem cells*. Evan Snyder of Harvard Medical School originally isolated neural stem cells from the cerebellum of newborn mice. (The cerebellum is a large structure at the back of the brain that, like the striatum, is concerned with the regulation of movement.) Unlike the dopamine cells we use for transplantation, Snyder's neural stem cells divided in tissue culture, so once he had isolated them he had what was basically a limitless supply. When Snyder grafted his stem cells back into mice, they developed into all the different cell types found in the nervous system, including nerve cells, the supporting glial cells, and even the cells of peripheral nerves.

More recently, Snyder isolated neural stem cells from the brain of human fetuses. He and I collaborated on a set of experiments in which we injected the cells into the brain ventricles of monkey fetuses. What happened was that the cells migrated out of the ventricles and turned into a wide range of cell types throughout the brain. Snyder is researching the possible use of these cells to treat children (or human fetuses) with inherited disorders of the nervous system where certain key enzymes are missing; the thought is that the transplanted cells might provide enough of the enzyme to prevent the disastrous clinical consequences of such disorders—mental retardation or death.

Snyder's neural stem cells might be useful in Parkinson's disease also, but an even more interesting possibility is to make stem cells turn into dopamine cells before transplanting them. Neuroscientist Ron McKay, of the National Institute of Neurological Disorders and Stroke, has made some exciting progress in this field. Rather than using Snyder's neural stem cells as the starting point, McKay uses embryonic stem cells. By treating the cells with a sequence of specific growth factors, McKay can get the cells to develop, first into neural stem cells, and then into dopamine cells. McKay has been working with cells from mice, but the Geron Corporation, which has licensed the patent on human embryonic stem cells from the University of Wisconsin, is believed to be working on the development of human dopamine cells by a similar procedure.

These developments, though promising, don't entirely get away from the ethical concerns about the use of human fetal tissue. Even though the supply of cells derived by one of these techniques would be essentially limitless, they would still trace their ancestry back to a human embryo or an aborted human fetus. Some people might consider the cells morally tainted for that reason. On the other hand, if these tissues were not used for research and human therapy, all would have been discarded.

There has also been considerable political debate about stem cell research. During the 2000 presidential campaign, George W. Bush said that, if elected, he would ban all federal funding for human stem-cell research. After he became president, Bush found himself caught between two diametrically opposed camps: moral conservatives, who wanted all work of this kind halted, and patients' advocates and the biomedical community, who wanted it to go forward. Not all conservatives were opposed to stem-cell research, however. Former First Lady Nancy Reagan, for example, whose husband Ronald Reagan is in the advanced stages of Alzheimer's disease, made it known that she wanted President Bush to support stem-cell research. Senator Orrin Hatch of Utah said that "embryonic stem cell research is the pro-life position."

In August of 2001 Bush announced his decision: federal funding would be permitted for research on embryonic stem–cell lines that already exist, but not for any lines that might be derived from human embryos in the future. While this compromise decision seemed to meet with public approval, we researchers are concerned that the existing cells may not prove useful, and some may not be stem cells at all. Only a small number of stem-cell lines have been isolated so far: Bush claimed that there are sixty lines worldwide, but the real number of useful lines is probably far smaller than that—some think a dozen or even fewer. Researchers may have to isolate hundreds of lines to find the few that are really useful—that is, cells that will remain in an undifferentiated state in tissue culture until they are given a chemical or genetic command to become a cell type that can be used to treat a specific disease. Bush's statement was at least a statement that research should proceed. It may have to be amended to get cells that are useful for human therapy.

It may ultimately be possible to get away from the use of cells derived from fetuses or embryos altogether. It's known that brain cells—both the nerve cells and the supporting glial cells—undergo a slow turnover even in adult animals, including humans; some cells die and are replaced by new ones. There must be a source for the new cells—these could be either neural stem cells that survive into adulthood, or they could be *progenitor cells*. The latter are cells that are one step beyond stem cells—they give rise not to a whole organ but to a specific class of cell such as neurons or glial cells.

In 1999, a group led by neuroscientist Fred Gage at the Salk Institute in San Diego reported that they had succeeded in extracting cells from the optic nerves of adult rats that, once treated with a series of growth factors, acquired the characteristics of neural stem cells. If the cells had been left in the optic nerve, they would have developed into glial cells, but the growth factors to which they were exposed caused them to backtrack in the developmental process and regain the ability to form any kind of cell in the nervous system.

This finding awakens us to an interesting possibility—that one might be able to obtain the cells needed to treat Parkinson's disease from the very patient who is to receive the transplant. This would be a tremendously exciting development, because it would solve three problems at one stroke: the difficulty of obtaining fetal tissue, the risk of immunological rejection (since they would be the patient's own cells), and the risk of transmitting diseases such as AIDS or hepatitis.

Where would one take the cells from? Not from the optic nerve, certainly. One possibility would be to take a biopsy from the lining of the cerebral ventricles. This is the region where new nerve cells and glial cells are born, and is thus a potential source of neural stem cells or progenitor cells. In fact, a Canadian group—Brent Reynolds and Samuel Weiss at the University of Calgary—have shown that it is possible to obtain stem cells from this source. The drawback in a clinical context would be that taking any kind of brain biopsy carries significant risk—of causing a stroke, for example.

What if one could reprogram cells even more radically? Recent work has shown that stem cells from one organ can, in certain circumstances, be persuaded to develop into cell types from a completely different organ. The more we find out, the more this kind of radical reprogramming seems feasible. Ultimately it may become possible to take, say, a muscle biopsy, isolate the muscle stem cells, and turn them into dopamine neurons. It's just a matter of learning the secret language of development.

~

Of course, one thing would beat repairing the brain, and that is protecting it from damage in the first place. Here, too, recent discoveries point to possible strategies—techniques for warding off Parkinson's disease or at least for preventing its progression from a minor irritant to a life-threatening disability.

Here's an example—a set of intriguing experiments done by Jeffrey Kordower, the neuroscientist at Rush-Presbyterian–St. Luke's Medical Center in Chicago who has been a key collaborator of Warren Olanow in his transplant studies. Kordower's experiments involved GDNF, the same growth factor that was used by Barry Hoffer to create a new nigrostriatal pathway in rats. Kordower wanted to see whether GDNF had any "rejuvenating" effect on the dopamine cells in the substantia nigra of aged monkeys. Rather than inject GDNF itself, which would be a short-lived treatment, he injected a virus carrying the gene for GDNF. The virus was taken up by cells near the injection site and these cells then made and secreted quite high levels of GDNF over a considerable period of time. He found that the production of dopamine in these aged monkeys did in fact increase.

Encouraged by this result, Kordower did an even more interesting experiment. He gave monkeys MPTP, the drug that kills dopamine cells in the substantia nigra. Then, within a day or two, he injected the virus with the GDNF gene. He found that the virus injection actually prevented the ill effects of the MPTP—the dopamine cells in the monkeys' substantia nigra survived. In other monkeys he waited a month after the MPTP treatment before injecting the virus—by which time the dopamine cells were already dead. In these animals the virus injection provided no benefit. In other words, the GDNF seems to protect dopamine cells from the toxic effects of MPTP, but it is not able to regenerate the cells when they have already been destroyed.

The patent on GDNF is controlled by the biotech company Amgen, and this company sponsored a clinical trial of GDNF in Parkinson's disease. The patients had a pump installed in the lateral ventricle on one side of their brain, and GDNF (the hormone itself, not the gene) was infused slowly over a period of months. I understand that the patients did not seem to benefit from this treatment. Even so, the ability of growth factors to retard the progression of Parkinson's disease is an avenue that deserves continued exploration.

Another way that Parkinson's disease might eventually be prevented is through the study of genes that cause or contribute to the disease. I've already mentioned that Parkinson's disease can run in families. The notion that this familial pattern is caused by specific disease-causing genes has been the longtime obsession of Roger Duvoisin of the University of Medicine and Dentistry of New Jersey. In the early 1990s Duvoisin finally came up with evidence to support his hypothesis. Aided by a multinational team of collaborators, Duvoisin studied a large family that originated in the town of Contursi in southern Italy, but whose members are now widely scattered around the world. Duvoisin reconstructed a family tree with 592 individuals, of whom no less than sixty have (or had) Parkinson's disease. Duvoisin's team, which included a contingent of molecular geneticists from the Human Genome Research Institute led by Mihael Polymeropoulos, went on to compare the DNA of family members with and without the disease. They found that the affected family members all possessed a mutation in a single gene, known as the α-synuclein gene. The gene produces a protein, α-synuclein, which is normally present in synapses in the brain but whose exact function is unknown.

This finding doesn't mean that the same mutant gene is responsible for most cases of Parkinson's disease—far from it. However, Duvoisin's discovery could still be of broad significance. It turns out that most people with Parkinson's disease, even though they don't have a mutant α-synuclein gene, nevertheless have abnormal accumulations of the α-synuclein protein in their dopamine cells. It is therefore possible that this protein is a key player in the destruction of dopamine cells, and that it can play this role either when it itself is abnormal or when it is triggered to accumulate by other factors. Interestingly, abnormal accumulations of fragments of α-synuclein are also seen in the brains of patients with that other great scourge of the elderly—Alzheimer's disease.

In 1998 a second gene causing Parkinson's disease was identified by a Japanese group, led by Nobuyoshi Shimizu of Keio University and Yoshikuno Mizuno of Juntendo University. This gene causes a rare juvenile form of the disease that can begin in the teenage years. The gene codes for a protein that has been called *parkin*, which is normally present in the dopamine neurons of the substantia nigra and helps these cells clean up damaged proteins. When parkin is genetically abnormal, fragments of damaged proteins accumulate and may damage the neurons.

It seems likely that further work on α-synuclein, parkin, and other genes will reveal the web of metabolic processes that lead to the death of nerve cells in Parkinson's disease. This in turn will open the door to the development of drugs specifically designed to block the process of degeneration. What is more, it seems likely that many of the elements of this metabolic web will turn out to be involved in other neurodegenerative diseases too, such as Alzheimer's and Huntington's disease. It's reasonable to hope that, with continued research, all these terrible diseases can be eliminated.

CHAPTER 21

Was It Worth It?

In July 1999, three months after we announced the initial results of the double-blind study, we held a reunion party for the participants. The party was held at Columbia College of Physicians and Surgeons, where the patients had gone for their five-day checkups. Most of the participants attended, along with their spouses or other family members; the exceptions were patients who would have had to travel a long way, such as James Ross and Sid Howard. Even the two patients who died were represented: Mildred Timmons by a nephew and Eugene Weiner by his wife.

Part of the purpose of the reunion was to inform the patients of the one-year results, so Stan Fahn, David Eidelberg, and I gave presentations similar to those we had given at the Toronto meeting. But the other purpose was simply to allow genuine social interactions in a study that had been, psychologically, a bit stressful.

My role in particular had been a difficult one. Because I knew which patients had received the real transplants and which the shams, I had not been able to talk freely with the patients during their first year after surgery. Even after that, I did not get to know many of them very well since they went to Columbia rather than Denver for their checkups. So it was with particular interest that I went from group to group during the party, asking the patients about their health and getting both their and their families' feedback on what the total experience of participating in the study had been like.

The first thought that came into my head as I walked around the room was "We haven't cured Parkinson's disease." Most of the participants were still easily recognizable as suffering from it in tremor, slowness of movement, dyskinesia, or other symptoms. Of these, some had not received a real transplant—Jack Celnik was an example. However, even among those who had received real transplants two or even three years previously, there were some with significant disability. One transplant recipient, Charles Frinton, was so dyskinetic that he had to lie down on the floor for a period of time.

Still, I also saw some dramatic success stories. Gregory Bennett was very obviously Parkinsonian before his surgery, but at the party he appeared quite normal. Emily Mason also looked normal, at least at the beginning of the party. Actually, one could have mistaken her for a member of the staff, because she made herself useful in a variety of ways, such as helping the arriving patients and family members to find their lapel badges. Toward the end of the afternoon, however, she became a bit dyskinetic, so her status as a patient was more obvious.

Although the results we presented were not uniformly positive, and the evidence in the room itself confirmed this, the feelings of the participants themselves were upbeat. In part this was because many of the patients felt that they had improved to some degree, even if they were not cured. In addition, most of the patients felt that their participation in the study had been a rewarding experience because they had done something positive and contributed to a meaningful study. The atmosphere was thus not one of recrimination or regret, but of celebration.

At the reunion party, and later, I found out how the various participants have fared since the end of their initial year in the study. Emily Mason has continued to do well. "Now I'm 'off' for about one and a half hours per day," she says, "but it's a minor 'off' compared with what it used to be. I may not be able to get to the door or phone in time, but I'm still mobile. In the morning I wake up 'on,' which is a wonderful change. Before, I had to take the medication and I'd be rigid for an hour before I could move. Now I get up and take a shower right away. I'm not totally 'on,' but I'm 'on' enough to get around. I do get dyskinesias, but usually it's because I overshoot on the Sinemet. I take it now pretty much by how I feel, which I never could have done before because my body was giving me wrong information. Now I can feel what my body needs." The most recent time I saw Mason was in September 2000. She and Giles had just done some hiking in the Rocky Mountains. She showed me pictures of herself crossing a stream by stepping from rock to rock. "I couldn't have done this before the transplant," she said.

Gregory Bennett also continues to do well. His ability to enjoy social occasions had increased greatly. "There may be a few times when his tremor acts up," says Pauline, "but it will be for a short period rather than for the whole evening." All in all, in spite of the delays and the problems with the sham surgery, both Gregory and Pauline are pleased with the outcome. "I have negative views about some of it," says Pauline, "but the end result for Gregory has been tremendously good, so I would do it again."

Sid Howard is now over eighty and continues in reasonable health, although between his heart problems and his Parkinson's disease he is in frail condition. Actions that most of us would complete in a few seconds, like descending a flight of stairs or putting together a sentence or two, can take him many minutes. But he insists on doing everything for himself, and he usually gets it done in the end. He has acquired a recumbent tricycle, which he rides down to the beach and back—his main form of exercise. In spite of his measured pace, he has taken a few tumbles from the bike, cracking a couple of ribs on one recent occasion. Asked whether he feels happy or unhappy about his participation in the study, he says, "Unquestionably happy. The only thing I regret is that I can't get back to playing tennis."

Jack Celnik, who was in the sham-operated older group, has elected not to have makeup surgery. He has experienced some deterioration in his condition since the time when he entered the study, and his overall feelings about the study are of disappointment and a certain degree of resentment. "I thought it could have been done with a little more finesse," he says. "A little bit more of the patients being involved in what was going on."

James Ross continues much as before. He is somewhat improved while in the "off" state, but he continues to experience dyskinesias which are severe at times. He has also experienced some cognitive problems, memory impairment in particular. But he continues to work on his mathematics and to play his favorite game, Go. "If I have the dyskinesia I'm a bit distracted and I may not play my best. But if I'm in a good state I play as well as ever. In fact, I think I've improved." Asked about his overall take on the fetal-cell transplants, he says, "The brain is mighty complicated, and anyone who thinks they know what's going to happen with this kind of thing is—confident."

Jacqueline Winterkorn is doing very well. She still notes slight improvements in her condition from time to time, over and above what she experienced during the first year after her transplant. She continues her medical practice on Long Island and her teaching at New York Hospital, and she and her long-lost boyfriend David— now her husband—enjoy life in rural Connecticut. Winterkorn has become an increasingly active member of her local Congregational Church.

The party did not mark the end of the study, of course. The participants still come to Columbia twice a year for checkups, and will continue to do so as long as the funds are available. In the spring of 2000, and again in 2001, we had further data-analysis sessions at Columbia, and the results were good. The younger group of transplant recipients had maintained their improvement and in some respects improved further over their one-year results. The sham-operated younger patients who received makeup surgery also did well as a group, improving over their preoperative state. The older transplant recipients, who had not benefited at one year after surgery, were better off two years after surgery, though still not as greatly improved as the younger patients.

Importantly, we have also figured out how to deal with the younger transplant patients who had had persistent dyskinesias even after reducing or even stopping all L-dopa. Paul Greene found he could block the dyskinesias by giving a drug called metyrosine. This drug stops the brain and the transplant from making L-dopa. Of course, that slows the patient down. It turns out, however, that adding back some L-dopa by mouth can help restore balance. Three of the patients with dyskinesias have needed to have stimulating electrodes implanted into an area called the globus pallidus, a procedure that is commonly done in patients with severe L-dopa–induced dyskinesias.

While I continue to work on the fetal-cell transplant surgery from a medical point of view, I am reminded from time to time that there are still political battles to be fought. One of them took place in Colorado. In April 2000, Representative Mark Paschall introduced a bill in the state legislature that would ban the sale or donation of fetal tissue for medical research within the state. "The trafficking of baby parts is so detestable and appalling that people, both pro-life and pro-choice, can scarcely believe that such a thing is happening right here in Colorado," wrote Paschall in a press release. Paschall suggested that the abortion clinics were selling fetal tissue (in contravention of federal law), but were concealing the payments as "reimbursement of expenses."

The fact is, none of the clinics where University of Colorado researchers obtain fetal tissue make a profit on the transactions. They do receive reimbursement of expenses, but that is exactly what they are—expenses. For example, a research group may need to use a room in the clinic for tissue processing, storage of supplies, and so on. The clinics are reimbursed at a fair rate for the use of this space.

On Monday, April 10, the House Health, Environments, Welfare and Institutions Committee held hearings on the bill. I went to testify against it. I explained that we currently had ten people with Parkinson's disease who were awaiting fetal-tissue transplants. I told the committee that only one in ten abortions yielded usable tissue, and that it took tissue from four fetuses to treat one patient, so that considerable numbers of donations are necessary. And I stressed that the University of Colorado does not buy human fetal tissue from anyone.

More powerful than my own testimony was that of two of my Denver-area patients, Don Nelson and Robert Majzler. Both Don and Carolyn Nelson came to the hearings, and Carolyn made the statement, partly because Don's voice is soft and indistinct, and partly to share her perspective as a caregiver. After briefly describing the history of Don's illness and the transplant operations, she listed some of the things Don has been able to do since receiving his two transplants, and emphasized his ability to become involved in social activities once again.

"We were told from the very beginning that this was not a cure," she went on, "however, our quality of life was so positively improved that we feel it was worth the risk of putting his life on the line for research to help not only himself but many others with Parkinson's, as well as other diseases. We hope this will not all end in vain. Please do not allow Don's efforts, along with the fetal tissue, to end up in a Dumpster as hazardous waste."

Robert Majzler remains a star patient—he isn't cured, but he is dramatically improved. When I saw him looking healthy and confident at the hearings, it was hard to believe that this was the same man I had examined a decade earlier—a man so disabled that I had to dress him myself and walk him out of the hospital—backwards!

After giving a brief background about himself, Majzler said:

> I was diagnosed at age forty-four. . . . My two sons were born after I was diagnosed. While they were still young, my condition worsened. I could no longer work. I was legally disabled. I could not get out of bed or get dressed without help. Once in my chair in the living

room, I sat there all day, every day. I was a prisoner in my own body. Because my wife worked and was gone for ten hours a day, I came to depend on my oldest son for help with my personal care, even though he was only six or seven. Like many disabled people, when I was out in public, people stared at me and looked at me as though I was crazy or drunk. The quality of our family life deteriorated quickly.

Then came the possibility of the transplant surgery. I was looking for something that would improve my existence—not just for myself but for my children who needed and deserved a father in their lives. The surgery resulted in slow improvement over many months. The fetal cells took root in my brain and grew slowly. With the passing of years, the cells grew stronger and produced the chemical reactions that my own cells ceased to do on their own. Over time, I was able to move on my own more reliably. With the return of movement, I was able to reverse my dependency on my children. Better still, I was able to be a dad to my boys, coaching their soccer and basketball teams. I have never been able to return to my career, but I have been a real part of my children's lives.

To illustrate the tremendous change that the implant made, I'll relate a short incident that happened to me recently. I was working out at a local recreation center and ran into a man I knew years before the surgery. Initially he didn't recognize me and seemed totally astounded to see me after all these years. He was so pleased and quickly added, "The wonder is that you are still alive!" I replied that I wasn't really alive until I had the surgery. The transplant gave me a second shot at living.

The Nelsons and Robert Majzler made a great impression on the committee, as did another witness who spoke about the importance of fetal-tissue research in the field of diabetes. The other side was represented only by a self-described pair of "Christian activists" named Jo and Ken Scott who wore T-shirts with a picture of an aborted fetus on the front. "We see the moms go into clinics with their big bellies," said Jo, "and then we see the ice trucks pull up. It's devastating." This seemed based more on fantasy than observation: most abortions are performed on women long before they have "big bellies," and if ice trucks come to the clinics it is not for any reason connected with the fetal tissue.

After the hearing, even Representative Paschall showed some change of heart. "I acknowledge that under existing technology some of these patients can only be helped through the implantation of fetal tissue," he told the *Denver Post*. He modified his bill so that donation of fetal tissue (with fair reimbursement of expenses) was permitted and selling tissue for profit was banned. This would bring Colorado law into accordance with preexisting federal law. The committee approved the modified bill and sent it to the full House. It has since become law.

The 2000 presidential election was déjà vu all over again for me: it brought back the events of 1988—the Election Day surgery on Don Nelson, and the battle between an anti-abortion Texan by the name of Bush who might block research and a pro-choice Democrat who could be expected to support fetal-tissue research. The outcome appeared to be depressingly familiar: George W. Bush's election suggested that the human benefits of fetal-tissue and stem-cell research might have to be postponed.

Right after Bush's inauguration in January 2001, Parkinson patients and their advocacy groups joined with others to lobby President Bush and the Congress, insisting that politics not block the great opportunities of stem cell research. Bush got engaged in the discussion and announced his decision in a special TV address to the nation on August 9, 2001. Much to the relief of patients and researchers alike, he gave the okay for NIH to fund research on embryonic stem cells. He restricted the funding to cell lines that had already been developed, but at least he gave us the chance to start.

~

On September 11, 2001, two planes hit the World Trade Center towers, changing everyone's lives. One of the thousands of people in the first tower to be hit was a fifty-seven-year-old electrician named George Doeschner, who was working on the thirty-fourth floor. George has had Parkinson's disease since 1986. Most people with Parkinson's disease take L-dopa—without it, they are frozen in place. Yet George was not taking L-dopa. We had given him a fetal-cell transplant in January 1999, and a year later he was doing so well that he was able to stop taking the drug altogether. Now he had the test of his life.

He started walking down the thirty-three flights of stairs with the thousands of others who were evacuating the building. As the crowd in the stairway paused to let more people enter, he noticed a tremor start in his right hand, but he did not freeze. He kept up the pace of everyone else and made it down the thirty-three flights in fifteen minutes. As the stream of people got to the bottom, hundreds were being directed toward an exit where, as George saw, pieces of the building were crashing down. The crowd changed its direction, surged across a bridge to the Winter Garden and Financial Center and moved about a block west toward the Marina.

Then someone yelled, "It's coming down, it's coming down!" George did not see the collapsing building but knew what was happening from the cloud of dust that looked like ash from a volcano. He first saw it surging down a parallel street a block away and then coming down his street. He started to run. "It was like being in a movie; everyone was running away from the monster." After running four or five blocks to the north, the cloud of dust thinned, and George and his companions stopped.

He needed to call his family on Long Island to say he was okay. He kept going north, but all the pay phones had long lines. When he finally got to a phone to call his wife, the recorded message said "All circuits are busy, please try your call again later." He got through to his son Derek and told him to call his wife. It was 10 A.M., just over an hour since the plane hit his building. He kept walking to Thirty-fourth Street and Penn Station, a distance of about three miles. No trains were running, so he waited. Finally, trains came and he got home at about 2 P.M., happy to be alive.

While Parkinson patients who have trouble walking down the street are said to be able to jump out of the path of a speeding car, George Doeschner made it down thirty-three flights of stairs, ran several blocks to avoid being overtaken by a cloud of debris, then walked three miles to find a train to get home, all without L-dopa in his system for over eighteen months. His fetal dopamine cell transplant had kept him moving and had kept him alive.

The enormous gratification of making a profound difference in some people's lives makes me willing to continue the political struggle alongside the scientific struggle. We will continue the quest to understand Parkinson's disease so that we can make treatments better. Sometimes it all seems a little daunting, but I only have to meet with a patient to be reminded how urgently we need better treatments—not just for Parkinson's but for a long list of degenerative brain disorders. More than anything, the courage of the men and women who have volunteered for our studies gives me the courage to move forward.

In my mind, Don Nelson will always be a hero. Don has always been a doer, yet his disease took away his ability to do many things and thus struck at his core sense of self. Thirty-two years after Parkinson's disease first touched him, thirteen years after he volunteered for our first transplant surgery, and five years after his second operation, Don remains a man of action. Recently he wrote me a letter to tell me about the improvement in his physical condition since the second transplant, and in typical fashion illustrated his progress by listing his accomplishments in home and garden:

> To be brief, I built two water ponds in my yard, connected by a stream and waterfall, two fountains, and a pagoda, built two planters in the front yard, installed a

drip sprinkler system, and rebuilt my train layout about three times the original size. Then I spent several weekends helping my son build a new log house in the mountains. We placed stone on the lower part of the house, painted the complete interior, installed all the interior doors and trim, and helped with the exterior stain finish.

As Don soldiers on, I have the same determination to push forward. For almost thirty years I have been seeking a solution to Parkinson's disease. We've shown we can repair the damaged brain. Now we need to translate that knowledge into a cure for this devastating illness.

Index

Index

Index

About the Authors

CURT FREED, M.D., is the director of the neurotransplant program at the University of Colorado. His cutting-edge treatments of Parkinson's sufferers have been featured in the *New York Times, Washington Post*, and *Wall Street Journal* as well as on *60 Minutes, Good Morning America*, and *20/20*. He lives in Denver, Colorado.

SIMON LEVAY, PH.D., a neuroscientist and science writer who has been on the faculty of Harvard Medical School and the Salk Institute for Biological Studies, is the author of six books, including *The Sexual Brain* and *Queer Science*. He lives in West Hollywood, California.

GAYLORD R